Game Based Organization Design

Game Based Organization Design

New Tools for Complex Organizational Systems

Jeroen van Bree

palgrave
macmillan

First published 2014 by
PALGRAVE MACMILLAN

Palgrave Macmillan in the UK is an imprint of Macmillan Publishers Limited, registered in England, company number 785998, of Houndmills, Basingstoke, Hampshire RG21 6XS.

Palgrave Macmillan in the US is a division of St Martin's Press LLC, 175 Fifth Avenue, New York, NY 10010.

Palgrave Macmillan is the global academic imprint of the above companies and has companies and representatives throughout the world.

Palgrave® and Macmillan® are registered trademarks in the United States, the United Kingdom, Europe and other countries.

ISBN 978–1–137–35147–0

This book is printed on paper suitable for recycling and made from fully managed and sustained forest sources. Logging, pulping and manufacturing processes are expected to conform to the environmental regulations of the country of origin.

A catalogue record for this book is available from the British Library.

A catalog record for this book is available from the Library of Congress.

Typeset by MPS Limited, Chennai, India.

Contents

List of Figures and Tables

Figures

Tables

Acknowledgments

Even though the words you are about to read were hammered out by a lone man behind a computer screen, the ideas presented here were the product of many wonderful collaborations which I would like to acknowledge. The two women who have had the biggest influence on my research work over the past seven years are Annemieke Roobeek, Chair for Strategy & Transformation Management at Nyenrode Business Universiteit, and Marinka Copier, Associate Professor in Play Design at the Utrecht School of the Arts, the Netherlands. A second group of people who influenced my thinking on these subjects are clients, in particular Tim Koelman, Han Hendriks, Leonie Voragen, and Maarten Korz. Then there are my co-researchers and colleagues, who helped refine ideas and approaches by reflecting on them with me: Thijs Gaanderse, Steije de Lat, Janine Swaak, and Ruud Kosman. Finally, I would like to mention the financial support by YNNO, without which the research that underlies this book would not have been possible.

The author and publishers wish to thank the following for permission to reproduce copyright material: OTH Architects for the photo (2010) used as Figure 6.1.

Every effort has been made to trace rights holders, but if any have been inadvertently overlooked the publishers would be pleased to make the necessary arrangements at the first opportunity.

1
Introduction

In the spring of 2006, I was invited to attend a business seminar on video games. It sounded entertaining enough, so I went. I did not really consider myself a gamer. Sure, I had a PlayStation 2 at home, but it was mostly gathering dust. My nights of guiding Lara Croft through the jungle were behind me, let alone my days of International Karate on my Commodore 64. But as I sat in the audience and listened to a speaker talk about so-called Massively Multiplayer Online Games (MMOGs), something happened. I saw a fascinating glimpse of a world that had been invisible to me until then. Millions of people were apparently leading parallel lives in these virtual worlds of *Everquest*[1] and *World of Warcraft*.[2] They had even established an economic system that could rival a small nation. I felt I had seen the future. This needed to be studied. Organizations had to draw lessons from this. The vague idea of combining my work as a management consultant with research that had been in the back of my head for years took shape there and then.

In the course of the years that followed, I have explored different aspects of the question that was the starting point for my project: what is the managerial relevance of video games? A big part of my journey has been looking for connections between this research question and existing fields of academic enquiry. I presented papers at seminars for video-game scholars, at academic conferences about computer-supported collaboration, and at colloquia for organization scientists. Each of these occasions gave me new insights, but each of those communities also had their doubts about what to do with this subject. Video-game scholars had a deep-rooted dislike for anything to do with 'management'. Academics in the organization sciences sometimes discarded video games as 'Skinner boxes' that are based on simple, addictive mechanics and have nothing to add to the organizational landscape. However, over the course of the years I have seen the intersection of organizations and video games grow in relevance. The initial skepticism has given way to fascination and to new buzzwords such as 'gamification'. For me, it has been an intriguing journey to get to the bottom of the potential that video games have for organizations. I gradually discovered that these new forms of teamwork and leadership that were revealed to me in the spring of 2006 are built into these games by the way they are designed. That is one of the central ideas that I developed in the course of my research: exploring the potential that video-game design has to enrich organizational design.

It seems that the time has come for these ideas about games and organizations to take hold in the business world. I hope this book can be a constructive contribution to this development. I have made an effort to be as interdisciplinary as my resources would allow. And I am fully aware that an interdisciplinary research project is vulnerable to criticism from each of the disciplines that it draws from. That is a risk I have willingly taken, because I strongly believe a full understanding of the subject at hand is only possible if one works from exactly this intersection. That is what I have tried to do, and it has made my journey so much more interesting.

This book is based on my PhD thesis,[3] and the fruit of work that took place over seven years. But we are not there yet. I consider this book to be an intermediate result, a stopover on my journey. I have set a course, explored the terrain, and am now ready to make some serious progress. I hope some of you will join me.

1.1   The research that forms the basis of this book

To increase the accessibility of the text, I have decided not to include an extensive description of the research methodology. I refer the interested reader to the PhD thesis itself.[4] What I will do in this section is give a brief overview of the core of the fieldwork that was conducted.

The fieldwork consisted of two action research projects that formed two cycles in which the theoretical ideas developed during the first part of the study were applied, evaluated, adjusted, and then re-applied. The first project was the Elective Care Center (ECC)[5] (described in

Chapter 4) and the second project was We Beat The Mountain (WBTM)[6] (described in Chapter 5). Both case studies were conducted using a 'research-oriented action research' approach (RO-AR).[7] The primary reason for the action orientation of this research is that it was concerned with a theoretical notion on which too little work had been done to formulate definitive hypotheses. To the best of my knowledge, there were and are no instances of organizations applying the video-game design process as a tool for organizational design and strategy formulation. Therefore, an exploratory study was appropriate. Both case studies should be considered testing grounds for applying a new type of organizational intervention, modeled after the video-game design process. The focus in both cases lay with evaluating the process of the interventions rather than the outcomes. My co-researchers and I performed the interventions ourselves, based on our theoretical insights. We followed the RO-AR cyclical process of 'extending theory-to-action-to-critical reflection-to-developing theory'.[8]

To counteract some of the problems of the action-oriented approach both on the practical level (performing and intervention and observing its effects) and the level of research rigor, an external researcher was invited to join the team. He had had no prior involvement with this work. His primary task was to make observations during the workshops, which involved a written report as well as taking photographs and recording short movies. The amount of information he received beforehand was more or less comparable to the information that the workshop participants received. Next to the observations recorded by the external researcher, other sources of data were interviews conducted by the author,[9] an online questionnaire, reflections by the research team, feedback sessions with the client team, and documents about the organizations in question and about the interventions.

1.2 Extension of the research

Since the time when I finished the first draft of my work in the summer of 2012, my thinking on the subjects discussed in it has continued to evolve. That shift in thinking is reflected in this book. There are two main reasons for it, one theoretical and the other practical.

First of all, I found that reengaging with the text and its subject matter led to a reappraisal of some of the things I had grown to take as assumptions in the course of my research work. I found I needed to challenge some of these assumptions or I needed to elaborate on the thinking that lay at its foundation. Especially the latter was an important part of the writing process for this book. In that sense, it should be much more accessible to the uninitiated reader than was my original text. Challenging my assumptions was also caused by being exposed to new sources that resonated with my ideas. Some of these sources are very recent, because work on the intersection of organizations and games continues at an accelerating pace. There were also some older sources that I had previously overlooked. Especially with regard to the recent work that is being published, I can of course make no guarantees that I am being complete. I integrated new sources into the text until the very last week of writing, but I am convinced that as I type these words there are researchers somewhere doing valuable work that is closely related to the themes discussed in this book. Let us keep the conversation going.

There have also been additions to the text that stem from practice. After the two action research projects that I discussed earlier in this section were completed, I began applying the ideas developed in this research more broadly, in consultancy projects over the past year. These projects were not accompanied by data collection and analysis in the manner of an RO-AR project, but they were nevertheless sources of valuable new insights based on reflections together with colleagues and clients. I have included examples from these projects in this text as well, most notably in the final chapter.

With these new insights, both from theory and from practice, I have taken the ideas presented in my PhD thesis a step further. In most cases, this has led to a refinement and strengthening of those ideas. In other cases – especially in the final chapter of the book – it could be considered an extension or conceptual extrapolation, which may still need further empirical evaluation. But I believe this broadening was vital to form a bridge to practitioners and to secure the connections not just to organizational design but also to the study of organizational strategy.

1.3   How to read this book

As I stated before in this introduction, I have tried to make this book as accessible as possible. I see two general ways in which the reader can approach it. The first is to read it as an introduction to a number of new ideas that revolve around the application of games, play, and rules to organizational endeavors. The chapters each deal with one core concept and in that sense can be read independently, although you will encounter quite a few cross-references to other chapters. This is unavoidable because the ideas presented progressively build on those of previous chapters. The notes supply plenty of material for further reading, for those who want to dive further into the subjects covered.

The second way to read this book is as a guide for a new approach to organizational design and strategy formulation. It describes a new set of tools for managers involved in those activities. If you are reading the book with that purpose, I recommend you read it from start to finish. Alternatively, if you are short of time, you can focus on the final two chapters. But that will require either substantial prior knowledge of games and play, or the willingness to refer frequently to previous chapters.

As a guide to the reader, the remainder of this introduction will give an outline of the chapters to follow.

1.4 Chapter outline

Chapter 2 describes how organizations are increasingly embedded in complex systems and how organizational leaders are struggling to address this complexity. Traditional modeling approaches to get a handle on those systems – such as systems dynamics and operations research – are briefly described, with a focus on their shortcomings when applied to social systems. A tendency to study successful organizations and to attempt to derive best practices is regarded as problematic because of the way these social mechanisms for success are tied to a specific organizational system. The chapter concludes with the identification of a gap between the complexity of the current organizational landscape on the one hand and the tools available for describing, understanding, and reconfiguring organizational systems on the other.

Chapter 3 starts with a brief history of video games and goes on to focus on a specific genre, Massively Multiplayer Online Games or MMOGs. The history of MMOGs is discussed as well as an examination of one of those games by the author. The attraction from a business perspective of the skills and behaviors witnessed in MMOGs is explored, as well as the barriers that exist to transferring these to organizations. Gamification is another attempt to introduce game elements in an organizational context. Criticism of gamification is examined, primarily in the context of the motivating factors of video games that this trend may not draw on. To conclude the discussion on games in an organizational context, the long-standing tradition of simulation gaming is described.

Chapter 4 describes the concept of play with the help of authors such as Johan Huizinga, Bernard Suits, and Brian Sutton-Smith. The basic qualities of play are covered, such as the fact that it is free and voluntary, and that it is carefully separated from the rest of life. The changing view of play from an organizational perspective is explored, as well as concrete attempts to link organizational play to learning and creativity. The concept of 'lusory space' is introduced as an isolated and temporary context in which organizational play can take place. A project is described in which lusory spaces are created, which leads to a number of practical recommendations.

In Chapter 5, the concept of rules is explored. The different ways of challenging the rules of games and other systems are explored in terms of the differences between being a spoilsport, cheating, and gaming the system. The latter is discussed in relation to the unintended consequences that often blight rule systems. The way in which games allow complex patterns to emerge from simple rules is examined in relation to approaches such as game theory and complex adaptive systems. The difference is then discussed between prescriptive rules (most common in organization theory), descriptive rules, and circumscriptive rules (as seen in games). The chapter closes with the description of a project in which organizational rules played a pivotal role.

The evolution of organizational design is discussed in Chapter 6, from viewing the organization as a formalized machine to a realization that design decisions are contingent on context, and to a current turn towards equipping managers with a unique mindset and approach to problem solving called a design attitude. The origin and essence of this design attitude are examined as well as their consequences for the appropriate organizational design process. The connection is then explored between this new form of organizational design and the video-game design process, characterized by prototyping involving the players, and the rule set as the object of design. This leads to a description of how this process could be applied to an organization, which is labeled game based organization design. This process is further explored in terms of the role of the organizational designer, the attributes of the design attitude it embodies, and the way it compares to similar approaches.

A brief introduction of strategy is given in Chapter 7, and the connection is shown between strategy formulation and organizational design. The traditional view of strategic management is then broadened by also looking at strategies for public and not-for-profit organizations, and by looking at strategy-making by other entities than just the corporate center. This extension fits the strategy-as-practice perspective. The organizational system is then discussed as a goal-driven configuration of environment, strategy, and structures and processes. The ways in which this organizational system can be redesigned in the context of game based organization design are explored. In addition, the boundary between the design process and the process of implementation and change management is examined. Finally, a description is given of a project in which game based organization design was used in the manner described in this chapter.

2
Systems

On September 16, 2008, President Bush sat down in the Roosevelt Room of the White House with Secretary of the Treasury Henry Paulson and Chairman of the Federal Reserve System Ben Bernanke. They discussed the events of the previous days. Lehman Brothers had filed for Chapter 11 bankruptcy protection the day before. In and of itself, the largest bankruptcy in US history was a dramatic event. But the unanticipated ripple effects were what really worried Paulson and Bernanke and led them to propose measures to the president that were previously unthinkable. The Lehman bankruptcy had led to the failure of a money-market fund that owned large sums of Lehman debt securities. This in turn had led to a general collapse of trust and a money-market run, which meant many large (non-financial) companies became anxious because they used this source of finance to fund their day-to-day operating expenses, such as paying their employees and their suppliers. The entire economy threatened to come to a standstill. After he had listened to the men, agreed to the proposed measures, and was concluding the meeting, President Bush put into words the precise perplexity which was puzzling many of us: 'Someday you guys are going to need to tell me how we ended up with a system like this.'[1]

2.1   The limits of mathematical models

There are many ways to interpret the events of that fateful week in September of 2008, but one way to look at them is as a clear demonstration of the need for business leaders and policy makers to understand the systems they are part of or aim to control. Or perhaps there is a step before that, which is the realization that your organization is embedded in larger systems. It is no longer possible to isolate yourself. Every organization to a lesser or larger extent has permeable boundaries. I realize that this is not a new insight. The basic idea of the organization as an open system was introduced in the 1960s by scholars such as Daniel Katz and Robert Kahn.[2] Around the turn of the century, under the influence of the rise of the Internet, the term 'network enterprise' became a popular label to attach to these open organizational systems. It was a term coined by sociologist Manuel Castells[3] to denote the organizational form he saw flourishing in various institutional and cultural contexts after the disintegration of the traditional, rational bureaucracy. Perhaps we can state that this development towards organizations with more permeable boundaries has accelerated in the past decades and that technological developments such as the rise of social media have further accentuated the reciprocal relationship an organization has with its environment. But that is not the issue I want address in this book. I am concerned here with the approaches and tools that organizations have at their disposal to get a handle on this dynamic and to determine their place in it in terms of organizational design and strategy. The dominant approach to gain an understanding of systems is still through mathematics. But the near-collapse of the financial system in September of 2008 showed that using mathematical models to get a handle on reality has severe limitations. Paradoxically, it demonstrated both the need for systems thinking and the limitations that common, mathematical approaches.

The optimistic belief that complex systems, even the world in its entirety, could be captured in a model reached its peak in the early 1970s. 1972 saw the publication by the Club of Rome of *The Limits to Growth*, a report that played and still plays an important role in the debate about population growth and resource depletion.[4] At the basis of the report lay a computer simulation called World3 that modeled aspects such as population and industrial growth. The World3 model can be traced back to the World1 and World2 models developed by Jay Forrester. An electrical engineer by training, Forrester had taken his expertise in systems thinking and simulation to the field of management in the 1950s. At MIT's Sloan School of Management, he established the field of Systems

Dynamics, a groundbreaking approach for modeling the dynamic behavior of systems.[5] Systems Dynamics has been a very influential field, which can be said to lie at the foundation of many current forms of systems thinking. In particular its holistic and endogenous view – that is, its attempt to include all relevant elements – allowed for a new way of approaching organizational challenges. At the core of systems dynamics lies a belief that systems can be captured in a model, which can then form the basis of a computer simulation to show how the system evolves over time. The World3 model developed for the Club of Rome revealed how fundamental new insights could be derived from such simulations. Within the context of the organizational landscape, the Systems Dynamics' use of simulations forms part of the broader field of operations research,[6] which uses immensely complex mathematical models to attack challenges such as the planning for complex mass transit systems and the supply chain optimization of oil, gas, and chemicals. Many of the complex systems that surround us and that form an essential part of our daily lives could not function without operations research and systems dynamics.

But there are limits. And the events of September 2008 showed those limits very clearly. Through financial engineering – closely related to operations research – complex and very profitable investment products were constructed. Ever fewer people could claim to understand the mechanics that lay at the base of these products. Somewhere along the line, the figures that these spreadsheets produced had started to replace reality. As it turns out, the financial experts in Wall Street, their bosses, their bosses' shareholders, and the institutions that were supposed to regulate them had lost sight of the human actors involved in the systems being modeled. Ultimately, it was the collapse of trust that almost brought the financial system to its knees. In other words, the human actors were the ones who threw a spanner into the machinery of the model. The financial system turned out to be also – or perhaps primarily – a social system.

The limitations of deterministic models when it comes to social systems had been recognized long before September of 2008. It was Russell Ackoff who, in the 1970s, began to criticize his own field of operations research.[7] The core of his assessment is worth repeating, because it does not seem to be fully comprehended, some forty years later. Ackoff placed operations research in a bygone era, which he called the machine age. In the machine age the dominant mode of thought was analysis and man believed he could understand the world around him by reducing it to its constituting elements and by uncovering cause-effect relationships. In other words, reductionism and determinism dominated our search for understanding. In the context of organizational life, this is the view inspired by engineering at the beginning of the twentieth century, regarding the organization as unified and formalized. The organization is a machine, to which general principles can be applied. It was the idea put forward by Frederick Winslow Taylor in his Scientific Management[8] and by Henri Fayol in his Administrative Theory.[9] But, as Ackoff aptly observed in 1979, when it comes to organizations the machine age is far behind us. We can no longer consider employees to be passive elements in a system. We have moved beyond the Fordist assembly line. Ackoff recognized that the members of the social system that is an organization have the ability to make a choice. In fact today, employees' ability to make these choices, respond to their dynamic environment and to keep thinking is generally considered an important ingredient of an effective organization. The knowledge worker – as identified by Ackoff's good friend, management scholar Peter Drucker – has become the focus of much management theory. But unfortunately, the knowledge worker refuses to be captured in a mathematical model.

2.2 Dealing with complexity

It is unfortunately not just the financial system that is in crisis. Other examples in the United States and Europe are the housing bubble, the unstable labor market, insecure pension systems, and the rising cost of healthcare. Complexity has now become the number one concern of CEOs worldwide, according to two recent studies. In a CEO study by KPMG International, 70 per cent of respondents said that increasing complexity was one of their biggest challenges. That same study showed that around half of the CEOs interviewed considered their actions to address complexity, such as business reorganization, to have been only moderately effective.[10] A CEO study by IBM showed similar results, with around two-thirds of respondents saying that their environments had become more volatile, more uncertain, and more complex to a large or very large extent. And in this study as well, CEOs report an inability to deal with this complexity. Less than half of those interviewed felt ready to handle the significant complexity ahead.[11] In his preface to the report, the then IBM President and CEO Samuel J. Palmisano identified the common denominator in their findings.

> We occupy a world that is connected on multiple dimensions, and at a deep level – a global system of systems. That means, among other things, that it is subject to systems-level failures, which require systems-level thinking about the effectiveness of its physical and digital infrastructures. It is this unprecedented level of interconnection and interdependency that underpins the most important findings in this report.[12]

So the big question that organizational leaders are facing is how to deal with the systemic nature of this complexity, both inside and outside their organization (that is, if making that distinction is still possible). From the studies by KPMG and IBM, an image emerges of CEOs who are realizing their toolbox is not up to the task. Management gurus such as Gary Hamel have further increased their worries by impressing upon business leaders the idea that they are still running their organizations based on management conventions from the early twentieth century.[13] Let me briefly discuss two compartments of the organization's toolbox that seem to be lacking. One concerns tools for understanding the organizational system, and the other tools for devising an appropriate course of action in response. Starting with the latter, one way to approach this is by looking closely at organizations that are successful. From the iconic *In Search of Excellence* – the Peters and Waterman study from 1982[14] – this has been a popular pastime of business journalists and organizational scholars alike (with the distinction between the two sometimes hard to make, save for the presence of regression tables). I will leave aside for now the problems of measuring an organization's success, especially in the long run, that may be associated with these studies, because there is another problem with this 'best practices approach' that I would like to address.

Perhaps as a sign of these confusing times, managers have lately become fascinated with organizations that seem to be acting contrary to the old business-school principles. How did Apple become arguably the most successful consumer electronics company of our time without undertaking any market research in the traditional sense[15]? How did Google become a multi-billion dollar company while its employees walk around their sunny campus in shorts and flip-flops, coming up with crazy projects like self-driving cars[16]? How did Ricardo Semler build a successful conglomerate by letting people manage themselves[17]? An additional source of fascination are non-traditional, loosely coupled organizations such as the open source and Wikipedia communities.[18] These are fascinating organizational endeavors which I highly recommend you study. However, the problem I have with mining other organizations for lessons is that these lessons are often drawn at the wrong level. That is to say, they are drawn at the superficial level of solutions. Managers see a specific solution or principle as the secret to a company's success and attempt to adopt that for their own organization. But installing the policy that employees can spend 20 percent of their time on projects of their own choosing will not necessarily promote innovation, as it does at Google. Within the context of the Google system, this policy has that effect. But to understand why, it would be necessary to understand the mechanisms of that system. Without having studied Google extensively, I can imagine it works because of a highly motivated workforce which is pushed to think beyond borders while at the same time holding each other's work to the highest possible standard. Or perhaps it works because Google can judge a very diverse portfolio of projects by one unifying goal, which is whether it will eventually extend their advertising platform. Whatever the reason may be, in Google's system this solution works. In the case of Apple, it becomes even more difficult to understand its success because of the obsessive secrecy that the company maintains. I would certainly not advise any company to forgo market research unless they have a singular design talent such as Jonathan Ive in their ranks, whose attention to detail is mimicked by the entire organization. Perhaps Apple's world dominance was born in that meeting of minds of Jobs and Ive, and will gradually come to an end with the passing of its idiosyncratic leader. The life and work of Steve Jobs is full of inspirational lessons, as is that of Ricardo Semler (luckily still with us), but there is no guarantee that their solutions will work if you transplant them to a different system, namely, to a different workforce, industry, or culture. Even if we were be able to prize the lid off completely and see the social mechanisms at work in producing these positive outcomes for Apple, Google, and Semco, organizational scholars would warn us that this can only be used to explain, not to predict.[19] What is more, it would be a step back to the machine age to assume that there is universality to these solutions. As designer and consultant Marty Neumeier puts it:

> The fact is, the experience of one company is not always transferable to another. You can't choose a solution from the 'solution shelf' as if you were buying a pair of pants from the ready-to-wear rack. In real life, you need to tailor your decisions to the unique challenge at hand, often working in dim light with incomplete measurements. In this situation, you can't DECIDE the way forward. You have to DESIGN the way forward.[20]

The subject of design is something we will return to later on in this book. But before the design of an appropriate solution can take place, there is a need for that other compartment in the toolbox, understanding your organizational system. I have stated before in this chapter that I believe that reducing complexity by analysis and mathematical modeling has serious shortcomings when applied to a social system such as an organization. An alternative to trying to reduce the inherent complexity of our contemporary organizational landscape is to attempt to embrace it. This holistic approach, the attempt to get a handle on the whole of a situation, has been a common feature of the many branches of systems thinking,[21] albeit often coupled with a (mathematical) modeling approach. As I will discuss in Chapter 6, the move away from reductionism and analysis and towards embracing complexity, chaos and messy, ill-defined problems is widespread among management theorists. It is an evident attempt to close the gap between the tools that managers have at their disposal and the dynamic of the system they are operating in. I would contend that we do not yet have the adequate tools for describing, understanding, and reconfiguring the complex organizational systems that currently predominate. Taking the acceleration towards permeable organizational boundaries as a given, I include the organization's environment in that conception of the organizational system. This book aims to be a contribution to closing that gap by looking outside organization and management theory for new insights and instruments that attempt to embrace complexity as a starting point. Specifically, I would like to draw your attention to the world of games.

3
Games

In November of 2011, Finnish video game developer Rovio announced that their game Angry Birds had been downloaded more than half a billion times.[1] To mark the occasion, the company released some statistics about the gameplay, which involves using a slingshot to launch wingless birds at pigs positioned inside increasingly complex constructions. One of the more astonishing facts was the number of birds that had been launched by players of the game since its release two years earlier reached 400 billion. In March of 2013, the number of downloads of Angry Birds (in all its different versions, including a Star Wars themed edition) stood at 1.7 billion,[2] so we can safely assume that the number of birds launched has tripled by now. These incredible numbers show how pervasive video games have become. In just a few decades they have evolved from an activity for reticent teenage boys to something that virtually everyone who owns a smartphone or tablet computer is exposed to. In fact, adult women now represent a bigger percentage of the gaming population than boys seventeen years old and younger.[3] How did we get to this point? And perhaps more interestingly, is it somehow possible to put that commitment to catapulting cartoon birds to use in an organizational context?

3.1 A brief history of video games

Although there were predecessors in scientific environments,[4] the first video game that reached the general public was Pong, developed by Allan Alcorn and Nolan Bushnell in the year that I was born,

1972. In essence, this was a rudimentary simulation of table tennis, with players moving their pads up and down to return the 'ball' to their opponent (who could be a human player or the computer). The instructions for the game were a testament to its simplicity and accessibility: 'Avoid missing ball for high score.' Pong was first marketed as an arcade game, that is, a game to be played on coin-operated machines in bars or gaming arcades. The game was a big success and is generally considered to be the launch of the video game industry. Bushnell co-founded Atari and, during the 1970s, established a quasi-monopoly of the gaming-arcade market. Around 1980, Japanese companies started to enter the video-game market, with by now classic games like Space Invaders[5] and Pac-Man.[6] In the bar that my grandparents owned stood one of those gaming machines, Space Panic.[7] As far as I can remember, this was my first exposure to video games, back in the early 1980s. In Space Panic, you had to dodge space monsters on different platforms connected by ladders. By digging a hole, luring an alien into it, and then covering it over, you could eliminate the vermin. This basic gameplay inspired a family of games, of which Lode Runner[8] is probably the best known one, and Space Panic may even have inspired the entire platform-game genre. It was a nerve-racking game, with oxygen running out slowly and space monsters moving faster and faster. But the best memory I have of Space Panic is of the days when the technician would visit my grandparents' bar. He was able to make my day by flipping a switch inside the machine that would set it to unlimited lives.

Atari was also successful in the home console market. Capitalizing on earlier devices that could be connected to a TV such as the Magnavox Odyssey and its own Home Pong console, Atari introduced its wildly successful home console, the Atari VCS, in 1977. This marked the entry of video games into the living room. Or at least into basements and attics, because video gaming remained a male-dominated hobby. In the early 1980s, video gaming also spread to Apple computers (in fact, Steve Jobs' first job was as a technician at Atari) and to now-extinct home computer platforms such as Commodore 64 and Commodore Amiga. In Europe especially, these home computer platforms became more popular than the game consoles in this period. My first home gaming console was one of those early personal computer platforms, the Sinclair ZX Spectrum (Figure 3.1), which even preceded the Commodore 64. By today's standards, this was of course a hopelessly primitive machine. It had 16 KB of internal memory and the software had to be loaded by means of a laborious process involving an external tape recorder. I would buy magazines that contained the actual source code for games and spend entire days typing in long strings of numbers and letters, because using hexadecimal code instead of the Spectrum's proprietary BASIC programming language would make the games run faster. More often than not, running the game would then result in a frozen screen, the loss of a day's work, and a sobbing teenager who sought solace in his Star Wars action figures.

At the end of the 1980s, the game consoles started taking over control of the market again. A veritable arms race took place that proceeded from Nintendo's NES (1986) to Sega's Mega Drive (1989) and finally to the revolutionary PlayStation introduced by Sony in 1994. The PlayStation broadened the market for video games considerably. More than 30 million PlayStations and 200 million games were sold in the first two years after its introduction.[9] I personally experienced the PlayStation as a quantum leap in gaming as well. After going through the Sinclair ZX Spectrum, Commodore 64 and Commodore Amiga in high school, I had lost interest in gaming during my

Figure 3.1   The Sinclair ZX Spectrum (photo: Bill Bertram)

university years. But during the 1996 Christmas holiday, one of my friends – who also grew up with the Commodore – suggested we go out and buy this new game machine, the PlayStation. We connected it to a TV, put the Tomb Raider[10] disc in, and marveled. Not having played video games for some years, we could not believe our eyes. This was a completely different experience, with depth, complexity, and a game character – Lara Croft – you could have somersaulting across the screen in all directions. For us, as for many others, it was the beginning of a renewed interest in this pastime from our youth.

The PlayStation was succeeded (PlayStation 2, PlayStation 3, and PlayStation 4) and challenged (Microsoft's Xbox, Xbox 360, and Xbox One) by consoles with increasing processing power, graphical sophistication and networking capabilities. The PlayStation and Xbox created a market for complex, large-scale video games with extremely high production values. Making these AAA-games (as they are called in industry parlance) requires teams of hundreds of developers and production budgets rivaling those of Hollywood movies. And although revenues of the U.S. video game industry increased from 3.3 billion dollars in 1996 to 10.3 billion dollars in 2002[11] (which put it on a par with the music and movie industries) its audience remained predominantly male and relatively young. This audience came to be known as 'hardcore gamers'.

Things started to change with the launch of three game-changing pieces of technology: Facebook in 2004, the Nintendo Wii in 2006 and Apple's iPhone in 2007. Of those three, only the Wii console was intended as a gaming platform. But Nintendo targeted a different audience than Sony and Microsoft. The Wii had a game catalog that lacked the AAA, action-oriented productions (partly because the hardware was not in the same class as the PlayStation 3 or Xbox 360) but instead offered casual, family oriented titles that made optimal use of the innovative Wii Remote, which allowed games to be controlled by the player's gestures. This new way of interacting with games was showcased in games such as Wii Sports, with players now standing side-by-side using their Wii Remotes as virtual tennis rackets instead of sitting alone on a sofa. Nintendo's strategy for broadening the gaming market proved to be extremely effective, with the Wii at one point outselling both the Xbox 360 and the PlayStation 3.[12] The Wii adjusted the image of the typical gamer, which could now include the whole family.

At the same time that the Nintendo Wii console was having its biggest success, two new gaming platforms emerged. One was the social networking site Facebook, founded in 2004 by Mark Zuckerberg and others and starting to increase popularity from 2009. One of Facebook's features is the ability to play a game on the site, alone or with your Facebook friends. The first game to really take off on the platform was FarmVille, released by Zynga in 2009. FarmVille was a game about managing your own virtual farm that made clever use of two game elements, for which it has also been criticized as being exploitative. The first was using the social networking features of Facebook to allow players to invite their friends who were not yet playing or to enlist their help in progressing through the game. The second element formed the basis for its so-called 'freemium' or 'free-to-play' model. This entailed not charging players anything up front but offering them the opportunity to progress quicker through the game by spending some money on extra 'farm coins', the game's currency. These coins could for instance be used to make the virtual crop on your farm grow faster. Zynga – which went public in December of 2011 – produced a series of games based on the FarmVille template that all relied on a small percentage (between 1 and 5 per cent) of their tens of millions of players to actually spend money on the game. The long-term success of Zynga remains uncertain, but the 'freemium' model has become dominant for online games on the Facebook platform and elsewhere, largely replacing the use of monthly subscription fees. The freemium model is not without its critics, though. One fundamental problem is that in some cases it becomes possible to buy your way ahead in the game, instead of earning it through putting in the hours. This is less of an issue for the players of Facebook games, but as the freemium model spreads to titles aimed at hardcore gamers – who generally spend a lot of time to achieve a certain status in their game – this touches at the core of the gameplay experience. That is the reason why the publisher of World of Tanks – currently one of the largest hardcore freemium games – recently decided to discontinue the sale of items that could be viewed as giving players an advantage in battle.[13] Another criticism of freemium games, and of the games by Zynga in particular, is that they are designed not with a fun gameplay experience in mind but with the aim of funneling the players to a point where they either give up out of frustration or spend money on the game to move ahead. For example, in the Zynga game Empires and Allies (recently taken offline) one of the tasks was to populate an island. As the buildings get bigger and the population grows, it takes more and more time for materials to be produced or houses to be built. It then becomes very tempting to spend some of your Empire Points on speeding up this process. These Empire Points are the game's currency, that you could purchase in the game or on the Zynga website. You could also use this currency to fill seats in your government building (if you didn't want to bother your Facebook friends with those requests) or even to auto-complete one of the game's missions. The criticism that Zynga's games push players towards spending money is not entirely fair, because games that rely on monthly subscriptions (such as many MMOGs) can also be said to make players spend an inordinate amount of time on menial tasks, which are not all built in to contribute to a fun gameplay experience.

The second new gaming platform that has emerged in the past years is the mobile phone and tablet computer. Apple's iPhone (released in 2007) and iPad (2010) were extremely successful in redefining what mobile computing means. Somewhat to the surprise of Apple, these devices also turned out to be very successful as gaming platforms. Which leads us back to Angry Birds, perhaps the best example of the mobile gaming experience. In 2012, the 157 million gamers in the United States spent half their time on mobile games and games on social networks or other 'casual gaming' websites.[14] These mobile and social games do not require the investment in time or money of console games and rely less on fantasy and violence, which have made them appealing to a much larger audience. In the United States, the United Kingdom, and other European countries, around half of the population is now playing video games of some sort.[15] Figure 3.2 gives an impression of the growth of the video game industry over the past decades.

Along with this broadening of video games and the gamer demographic, the discourse surrounding games has also started to change. There is still a knee-jerk reaction in the media that relates video games to violent incidents such as school shootings. But increasingly, we have seen discussions about video games shift to their educational value and to the skills being acquired by playing them, such as learning by doing, embracing the right amount of risk and pattern identification.[16] To illustrate this new perspective on video

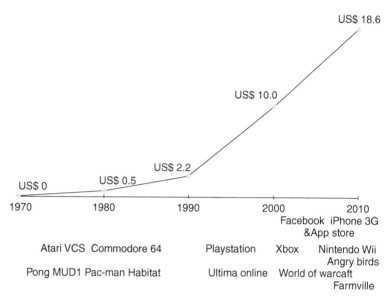

Figure 3.2   Timeline of major video game developments and revenues (in billions) of the US video game industry[17]

games, I will examine two recent examples of discussions about the utility of video games as they relate to the business world. The first revolves around so-called MMOGs and the second deals with a more recent trend called 'gamification'.

3.2   Massively multiplayer online games

There is one genre of video games that at first sight has the greatest link with organizational life: that of Massively Multiplayer Online Games, or MMOGs.[18] MMOGs are large-scale, complex online games, played while connected to a persistent virtual world where players can collaborate and form teams. The key to success in these games lies in increasing the power of one's avatar – the user's virtual representation – through the completion of tasks, alone or with others. This rise through the ranks requires a substantial time investment and commitment (regularly over twenty hours a week), making these games the complete opposite of the mobile and social games mentioned in the previous section. Tens of millions of people worldwide play these types of games.

3.2.1 The history of MMOGs

The history of MMOGs dates back to 1979, when Roy Trubshaw and Richard Bartle created the first text-based virtual world called MUD1 at Essex University. MUD stands for Multi-User Dungeon and refers back to the single-player adventure game DUNGEN or Dungeon that this system was based on. MUD1 was purely text-based. Users could move around virtual spaces by means of text commands. If another user was in the same space, they could communicate by means of a chat function. The original MUD1 spawned hundreds of followers in the course of the 1980s. Gradually, these MUDs or MOOs – the latter stands for MUD, object oriented – moved from a kill-or-be-killed game to more social environments. These social environments also attracted the attention of journalists and researchers. Especially the one called LambdaMOO became a focal point for academics because it was the first MUD to establish a substantial user base.

Sherry Turkle was one of the researchers using LambdaMOO for her fieldwork.[19] She focused on the fact that MUDs allowed users to inhabit a virtual space with others for the first time and she studied the social interactions and particularly the role-playing that took place there. Turkle talked about the idea of 'a decentered self that exists in many worlds and plays many roles at the same time'.[20] At the time, this was a new and exciting concept, because MUDs even preceded chat rooms, let alone text messaging and Facebook. Today, the context is a different one. And although MMOGs are still a way to experiment with different identities for a lot of people, life on the screen in many ways is much more tied to real life than it was fifteen years ago. Gameplay spills over into websites and chat rooms, and games are played with friends and family members.[21] Another author of note from the MUD period is Lynn Cherny, who performed an ethnographic and linguistic analysis of a LambdaMOO spin-off called ElseMOO.[22] Its lexicographic analyses are thorough and still worth a read, but this study of the ElseMOO community has lost most of its relevance because of one major development: current online games are truly massive and it is hard to translate the workings of the close-knit and relatively small MUD communities to millions of people playing the current generation of MMOGs. The changing context notwithstanding, these authors laid an important foundation for subsequent research on MMOGs and for their further development.

A significant missing link between the early MUDs and todays MMOGs was the creation of a graphical and persistent virtual world in which the action takes place. One of the first graphical virtual worlds was Habitat. It was released by Lucasfilm Games in 1986 on the Commodore 64 platform. One of the creators of Habitat, Chip Morningstar, was the one who introduced the term 'avatar' to denote the representation of the user in the virtual world.[23] Habitat was not in operation for very long and it took until the end of the century for virtual worlds to gain some momentum again. With the increase of computer power and the proliferation of the Internet a number of new social virtual worlds arose in the second half of the 1990s, of which Active Worlds (released in 1997) was the most prominent one. By now, the graphical sophistication of these worlds had increased to give the user the experience of using a radically different user interface. But still, because of the limited number of participants the phenomenon remained mainly of interest to researchers and computer scientists.

Things changed in the early 2000s. It was at that time that the move towards today's MMOGs started to gain momentum. A milestone in that respect was the introduction by Electronic Arts of Ultima Online in 1997. Besides showing all the characteristics of what we would today consider an MMOG, the most important thing that set Ultima Online apart was its commercial success. People were willing to pay $10 per month to play the game. When Ultima Online exceeded 100,000 subscribers within a year, the video game industry started to take serious notice. The genre of Massively Multiplayer Online Games or MMOGs was born. Ultima Online had a number of successors, of which EverQuest[24] was the most notable one. EverQuest topped Ultima Online's subscription numbers, reaching 300,000 subscribers in 2000. Developments in the field of MMOGs then started to move very rapidly. There were already games breaking the 1 million subscriber mark in Korea (Lineage, for example) but 2004 saw the introduction of a game by Blizzard Entertainment that topped all this: World of Warcraft.

For some time now, World of Warcraft has been the dominant game in this genre. As of October 2012, it had around 10 million subscribers worldwide.[25] These numbers are down from a peak of 12 million a few years earlier. Every subscriber pays €13 per month to access Blizzard Entertainment's online fantasy world of Azeroth, which gives an indication of the size of the consumer market for this genre of games. However, there is increasing pressure on the viability of this subscription-based model. The most important reason is the rise in MMOGs of the freemium model, mentioned in the previous section. This means players can play the game without a monthly subscription but must pay to get access to extra content or virtual items. This model has been the norm in Asia, but is now making headway in the West as well. Blizzard has now made World of Warcraft available for free until the player reaches level 20 (of a current maximum of 90).

The type of users populating these environments has broadened substantially over the years. Research by the Palo Alto Research Center and others showed the average age of World of Warcraft players in the United States to be thirty-two for women and twenty-nine for men. Europe shows similar numbers but Asian players are considerably younger.[26] The same study showed that a third of the players were female, more than half were in a relationship and 25 percent had children. Again, this was the picture for the United States and Europe, with decidedly more male and single Asian players. Some other interesting findings from this and similar surveys concern the average number of hours played per week (around twenty-two) and the fact that a majority of players either play the game with real-life friends or make new friends in real life through playing World of Warcraft. So even for MMOGs – which are considered hardcore games – the stereotype of the gamer as a lonely teenage boy does not hold, at least not in Europe and the United States.

In the second half of the last decade, these large-scale online games started to be taken seriously by academics and by the business community. Edward Castronova's 2001 paper titled 'Virtual Worlds: A first-hand account of market and society on the cyberian frontier' was groundbreaking in that respect. At the time of writing, Castronova was in the economics department of California State University. He stumbled upon the virtual world of Norrath (the name of the world in which the game EverQuest takes place) and became interested in its economy. Castronova showed that there was productive activity taking place inside this virtual world. As a matter of fact, 'you could earn a poverty-level wage, in real dollars, by "working" in the game'.[27] Journalist Julian Dibbell later devoted a book[28] to his yearlong attempt at making a living by doing just that: 'working' inside EverQuest. He also reported on the phenomenon known as 'gold farming': performing certain in-game activities repetitively (either automated or by employing low-wage workers) to produce goods that can be sold for real money.

3.2.2   An expedition into World of Warcraft

To better understand the appeal and the potential of MMOGs, I decided to organize my own expedition into World of Warcraft (WoW) in the fall of 2007. I had not previously played this game or any MMOG, so I was an absolute beginner. To be able to play the game I bought the boxed World of Warcraft software and took out a monthly subscription. I played the game over a period of approximately three months using characters in two different classes: first I played as a priest, who is in demand by groups for his healing capacity and needs to rely on those groups to advance in the game because his armor and weapons are not very powerful. When progress using the priest character became somewhat frustrating I decided to start again using a hunter, who is more suitable for solo play because of his superior fighting skills. He has longer-range weapons such as a crossbow and can even send his pet tiger to attack on his behalf. Using the hunter, I eventually progressed to level 22 (of 70, which was the maximum at that time). I experienced 'grouping' with other players to complete game 'quests', but did not reach the level of large-scale 'guilds' that go on 'raids'. There will be more about these terms in quotation marks later on.

The players of World of Warcraft are spread out over servers of a few thousand users each, tied to a specific region (United States, Europe, or Asia). Because this expedition took place on one server (a European server called Nordrassil), I also engaged in discussions on WoW-related websites such as WoW Insider that cover the community as a whole as well as the higher-level gameplay, which I did not experience.

So what did I learn during my expedition? One of the important aspects of these games that I experienced is that each player is by definition incomplete. Classes and races are designed in such a way that they are complementary.[29] Depending on the class you choose to play, it can be difficult to advance in the game by yourself. While it is possible in theory to reach the higher levels of World of Warcraft without the help of others, teaming up – or 'grouping', as it is called in the game – usually makes it a lot easier. And when a player has reached the higher levels (level 60 and up), the game becomes almost exclusively available to groups. This part of the game consists of so-called 'raids', missions where a large number of players combine forces to defeat a powerful enemy. These raids are carried out in groups of up to forty or fifty players where a number of very specific roles need to be filled, such as the tank, the DPS and the healer. The tank is the player who confronts the opponent head-on, the DPS (which stands for damage-per-second) is the player who tries to inflict as much damage as possible to the opponent, usually from a distance, and the healer makes sure everyone in the team stays healthy. But also earlier on in the game, quests are presented to players that can be too difficult to complete alone. Quests are tasks that – when completed – bring the player closer to a higher level and also deliver virtual money or goods that can be used in the game. The game system explicitly advises you to group for certain tasks and supplies a specific tool in the user interface to look for other players that want to complete the same quest.[30]

When collaborations between players become more permanent, they usually take the form of a 'guild'. A guild is a team structure supported by the game for ten or more players that decide to band together. Guild members are identified in the game by a tag above their name and they have access to private guild chat channels. Often, the guild life extends outside the game space with websites or even real-life meetings. Dmitri Williams of the University of Illinois did an extensive study of guild life in World of Warcraft together with researchers from the Palo Alto Research Center.[31] They distinguished between the casual or social guild and the raiding guild. The latter focuses on the necessity to group up in order to advance in the game, a dynamic that was also shown by other researchers.[32] The casual guild type shows that for many users it is also a goal in itself to socialize with others in the game. Guilds can have different degrees of formal organization. Using an analogy to group sports, guilds can range from a group of friends playing soccer in the park from time to time to a soccer team competing in a recreational league every weekend. The latter form obviously requires more coordination and leadership and in guild terms the guild master becomes an important role. He or she has to coordinate the guild activities, handle recruitment, negotiate guild mergers, and police disputes.

Another aspect of interest in MMOGs is the reliability of the signals that the avatar gives off about the user. The hairstyle, skin color, and even sex of the avatar do not say much about the user because these traits can be chosen freely. However, there are a number of other signals that are reliable. First of all, there is a visual difference between players of different classes and of different levels. The different classes signal certain unalterable traits, such as the fact that the priest class has the power to heal other players. The levels have an even stronger signaling quality, because of the large amount of time that needs to be invested to progress through the levels. Having an avatar with a high level signals a certain amount of experience and abilities (abilities are added more or less automatically as you gain a level). When my level 7 Priest character ran into a level 70 Hunter, after a week and a half of struggling through the first levels, I was in awe of the amount of time and effort this person must have put into the game. On closer inspection, one is even able to see to some extent which game tasks the other player has mastered (because of the clothing the avatar is wearing). With a bit more effort, other data about fellow players can also be obtained, such as their armor, weapons, professions and 'build'. A player's build shows the specific direction chosen in the development of a character; e.g., a priest can choose to focus on his healing ability or on his ability to do damage. All this data is objective and difficult to falsify. These signaling functions play an important role in quickly establishing groups for collaboration.[33] It seems that groups are formed largely on the basis of the experience and abilities of players. Elements such as age, gender, or location play a relatively limited role. Byron Reeves of Stanford University described this phenomenon as a meritocracy: teams are based around skills instead of social relations. His article (together with Thomas Malone and Tony O'Driscoll) in the May 2008 issue of the *Harvard Business Review* declared games such as World of Warcraft to be 'Leadership's Online Labs'. It was the peak of the interest of the business community in MMOGs. The phenomena that were witnessed in these games spoke to the imagination of many managers. When I explained the game and showed videos of the complex World of Warcraft gameplay to business audiences in those days, their reaction was often one of amazement and of wishing distributed teamwork would take place in a similar, well-oiled fashion in their organizations. In addition, I sometimes had people come up to me after a talk to express their relief that they finally understood what their son was doing behind his computer screen and that he was apparently learning something.

To understand the appeal of MMOGs to the business world, it is important to acknowledge the limitations of the technology for virtual teamwork that these games were often compared to. The technology for virtual teamwork is part of the field called Computer Supported Cooperative Work.

3.2.3   Interlude: Computer Supported Cooperative Work

The academic field of Computer Supported Cooperative Work (CSCW) has been characterized, since its inception in the mid-1980s, by a combination of an attempt to understand how people work in groups with efforts to support that work with information and communications technology (ICT). The field was born out of a need to fill the gap between single-user applications like word processing and large mainframe systems that served a whole organization.[34] The development of these new ICT systems was motivated by one or more of three underlying assumptions[35]:

- improving task performance of work groups;
- overcoming space and time constraints on groups;
- increasing access to information.

Technologies in the field of CSCW included video conferencing systems, e-mail, instant messaging, and virtual workspaces. Many of these collaboration technologies have moved from being isolated applications to being integrated with other tools or to being an almost invisible part of the general infrastructure that is available to users.

CSCW touches upon the topic of computer-mediated communication (CMC). Since its emergence, this new form of communication has been the subject of much academic interest. Researchers have been studying its effectiveness and its social characteristics, compared to other forms of interaction such as face-to-face meetings and telephone conferences. One of the best-known conceptual frameworks that is used to compare forms of interaction is the media richness theory by Richard Daft and Robert Lengel: 'Communication transactions that can overcome different frames of reference or clarify ambiguous issues to change understanding in a timely manner are considered rich. Communications that require a long time to enable understanding or that cannot overcome different perspectives are lower in richness.'[36] Daft and Lengel considered face-to-face to be the richest medium and impersonal written documents among the leanest. Although developed before the widespread use of CMC, the media richness theory has been used extensively for developing a fit between task and medium. The general idea put forth by Daft and Lengel is that a richer medium is needed for situations that are more equivocal (that is, ambiguous). Evaluations of the media richness theory have introduced a more nuanced look at the fit between the medium and the situation. Joseph Schmitz and Janet Fulk have argued that Daft and Lengel's media richness is too normative.[37] According to them, there is no such thing as an objective information richness for a specific medium. It is dependent on many factors such as the computer experience of the user, making it a much more subjective concept. Joseph Walther[38] has shown that social information can be exchanged through CMC, but that it just takes more time than through face-to-face contact. He even presented evidence of CMC being able to surpass face-to-face contact when it comes to social orientation based on phenomena such as the tendency to build stereotypical impressions of your counterpart based on an overreliance on minimal cues, and the possibility of optimized self-representation because of fewer cues and the possibility of editing one's comments. What these authors show is that there is a need to take a nuanced look at the fit between the medium and the situation. A high degree of equivocal or social information does not necessarily mean resorting to face-to-face contact.

Besides the amount of ambiguity that can be handled by a medium, another challenge in CSCW has been that of awareness: 'actors' taking heed of the context of their joint efforts'.[39] This term has been used in two, distinct ways:

- Awareness of the social context of work. An example of this is the early work on media spaces at the Xerox PARC lab,[40] where the need for supporting informal interactions was stressed.

- Awareness as 'practices through which actors tacitly and seamlessly align and integrate their distributed and yet interdependent activities'.[41] This notion is based in large part on the groundbreaking ethnographic studies into the work of collocated groups of air-traffic controllers by Harper, Hughes, and Shapiro[42] and those of controllers of the London underground by Christian Heath and Paul Luff.[43] These studies showed workers displaying and monitoring activities in an unobtrusive way, also termed consequential communication.

The need for consequential communication in a distributed situation can be met by more explicit mechanisms such as telling your colleagues what you are doing or feedthrough (observing changes in artifacts, such as documents).[44] It can also be partly replaced by peripheral participation ('overhearing' electronic discussions or chats) and broadcast communication.[45] Awareness of the social context of work has been a tougher nut to crack. It touches on the need for informal interactions, without which 'many collaborations would undoubtedly not occur and others would break up before becoming successful'.[46] Finding the 'virtual water cooler' as a natural place for informal communication remains one of the formidable challenges of supporting distributed work. The difficulties with ambiguous information and the lack of informal or consequential communication have put major limitations on the ICT support for distributed work. It will frequently not be possible to support all social aspects of the work patterns in the case of a distributed group. This is the basic frustration of CSCW research. It is what Mark Ackerman of the University of Michigan calls the social-technical gap: 'the great divide between what we know we must support socially and what we can support technically'.[47]

3.2.4 Transferring behaviors and skills

In contrasting the unfulfilled promise of computer-supported cooperative work with the potential for seamless online collaboration witnessed in MMOGs, some managers, journalists, and researchers saw a new set of technologies to be used for virtual teams in large enterprises. This was a perspective that seemed to remain limited to what happens on the screen, and viewed video games primarily as a new piece of technology that could somehow be integrated into business software.[48] But a closer inspection shows that the real power of video games runs much deeper than what we can witness on the screen. It may have much more to do with what happens *in front of* the screen: the impact of games on player behavior and on the skills they are acquiring through playing.

Is it possible then for the skills that these gamers are acquiring to be transferred to an organizational context? Are MMOGs indeed 'leadership's online labs'? There exists a barrier to the realization of this attractive vision and that is the sizable gap between video games and the typical corporate environment in terms of what learning scientist David Shaffer calls epistemic frames: 'collections of skills, knowledge, identities, values, and epistemology'.[49] Shaffer discusses epistemic frames in relation to what he terms 'epistemic games': games that teach young people necessary skills by assuming the epistemic frame of creative professionals. The epistemic frame of a game such as World of Warcraft is one where skill trumps personal attributes such as age or gender (mostly because the latter are invisible), where teamwork is an absolute necessity for progress, where leaders can become followers and vice versa, where every player has the tools to manage their own development and where a level of proficiency in the highly complex user interface is a barrier to entry. In no way does it resemble the epistemic frame of the typical corporate environment, where old structures dictate how leaders rise to the top, how much autonomy employees have and how information flows. We may of course nurse the hope that the corporation's epistemic frame is growing closer to that of World of Warcraft in these respects – which is certainly the case in companies such as Google and Semco, mentioned in the previous chapter – but we are not there yet. According to Shaffer's theory, a leadership skill acquired in the context of one epistemic frame (World of Warcraft) is not necessarily effective in the other (a corporation). Shaffer's point is further supported by a recent study by researchers at the University of Gothenburg,[50] who uncovered forms of collaboration in MMOGs that seem locally tied to a particular game. They take the position that it would be hyperbole to claim that these games foster collaborative skills *in general*. In other words, the transfer of these skills to a non-game environment is certainly not obvious. Perhaps an even more interesting question is whether gamers themselves even *want* to take the skills they learn and the epistemic frame they encounter in their game back to their working life. Many of the more optimistic authors assume they do, but Harald Warmelink of Delft University of Technology decided to actually ask online gamers about this. He found that some 'find the idea of comparing their online gaming community organizationally to their work organization strange or problematic'.[51]

It is important to keep these warnings about transfer in mind as we discuss a recent wave of interest in applying the power of video games, which has come to be known as 'gamification'.

3.3 Gamification and the potential of games to influence behavior

In the course of 2011, the term 'gamification' started to surface. It was a label attached to a collection of initiatives aimed at using games or game design elements in non-game contexts.[52] What set it apart from earlier uses of games for non-entertainment purposes – or more accurately: not purely for entertainment purposes – was that gamification aims to overlay reality with a permanent game layer. This is opposed to the more established use of (video) games for education and training, usually labeled 'simulation gaming' or 'serious gaming' (more about simulation gaming in the last section of this chapter). An important objective for using serious games is to learn, particularly in situations where learning is dangerous or where teaching is expensive. However, the aim of gamification is not to create a finite, simulated situation that is used as an instrument for learning, after which the lessons will have to be transferred to the context in which they are to be applied. Gamification aims to make a game of (a part of) reality. In itself this is a daunting endeavor, which possibly goes against some of the fundamental characteristics of play (more about that in the next chapter). But a bigger problem seems to be that many of those involved in gamification have a limited understanding of the core concepts of games.

Gamification started out with fairly innocent ideas such as the Chorewars application[53] that transforms household chores into an MMOG as well as the many examples in The Fun Theory initiative by Volkswagen,[54] which encouraged sustainable behavior (such as recycling bottles) by making it into a little game (the 'bottle bank arcade'). Another early example was Nike+,[55] which allowed runners equipped with Nike's app or wristband to keep better track of their progress in order to beat their own best or compete with friends. And there was the successful Foldit project,[56] which let participants play a game that helped scientists solve the complex problem of protein folding. The players of the Foldit project made some major contributions to developing a cure for AIDS.[57]

All these examples are creative ways of using game elements in a non-game context. They make use of game mechanics: the mechanisms (the terms 'mechanism' and 'mechanic' are used interchangeably in game design literature) that are behind a game and that evoke certain behavior patterns in its players. In these cases, game mechanics are used for the greater good of a healthier, cleaner world. But a more prevalent use of gamification is to increase engagement with online applications and websites. The emblematic, and one of the most successful examples of this is Foursquare: an application that allows you to score points by 'checking in' at different locations using your mobile phone. Even though Foursquare was not explicitly labeled a game, it is a perfect illustration of one of the simplest game mechanics: the player receives points for performing an activity, the points are added up, and the total is placed on a leaderboard. The player who comes out on top gets a reward. In the case of Foursquare this reward can be a virtual badge (the player can become the 'mayor' of a certain location) or a coupon for a free cup of coffee. This points-leaderboard-reward template has been used by companies such as Badgeville and Bunchball to offer solutions for 'gamifying' companies' online interactions with their customers or to boost the productivity of employees.

The gamification trend has received a fair bit of criticism from video-game designers and game scholars. Their criticism has centered on the contention that using isolated game elements does not constitute proper game design. According to many game designers, full-fledged games are based on more refined mechanisms, with points and badges being the least important part of games that do not give access to their full potential.[58] One of the most outspoken critics of gamification has been Ian Bogost, professor at Georgia Tech and co-founder of game studio Persuasive Games. He is best known in game circles for Cow Clicker, his satire of the Facebook game FarmVille (discussed earlier in this chapter). Bogost's frustration about gamification stems from his fear that the mysterious, magical, wonderful medium of games is snatched from his hands, trivialized, and then sold back to him in a watered-down version. This is a version with limited replay value, which diminishes the usefulness of gamification as a *sustainable* strategy for customer engagement or behavior change. This point was substantiated by the few academic studies so far into the effects of gamification. One of those studies was by researchers at the Delft University of Technology.[59] They found that game elements are able to influence behavior (in this case: reduce energy use) but that this behavior change turned out not to be sustainable in many cases. Another recent study of the use of virtual badges to increase customer engagement (a popular gamification technique) on a community sharing platform showed no significant increase in the users' activities.[60] Despite the lack of evidence for its effects, the use of gamification seems to be on the rise. There are new examples every week, with applications ranging from interacting with your bank to managing your love life. One research firm expects the gamification market to reach $2.8 billion in the United States by 2016.[61]

One of the problems of the simple points-leaderboard-badges game mechanic is that it does not tap into intrinsic motivation. When activities are not performed for their inherent enjoyment, they are said to be extrinsically motivated. Richard Ryan and Edward Deci distinguish between four types of extrinsic motivation in their Self-Determination Theory,[62] varying in the extent to which the regulation of the behavior has been internalized. The subtle differences between the two most internalized forms of regulation (identified regulation and integrated regulation) go beyond the scope of this discussion. However, it is useful to make a distinction between the two other forms of regulation:

- External regulation of behavior: these behaviors are performed to satisfy an external demand or reward.
- Introjected regulation of behavior: these behaviors are performed to avoid guilt or anxiety or to attain ego enhancements such as pride.

Many of the behaviors that gamification attempts to generate are performed to satisfy an external reward such as points and badges (external regulation) or to attain ego enhancements such as beating your friends' ranking on the leaderboard (introjected regulation), as depicted in Figure 3.3. This could be seen as a missed opportunity, because it has been extensively discussed in digital games research that video games supply their own reward.[63] They are intrinsically motivating.

One of the most quoted theories of intrinsic motivation in video games was put forward by Mark Lepper and Thomas Malone.[64] According to them, intrinsic motivation in video games is enhanced by:

- creating a balance between skills and challenges;
- stimulating the sensory and cognitive curiosity of users;
- providing a sense of control to the user;
- creating fantasy situations.

Many of these elements of intrinsic motivation can be found in other theories as well. The first and third elements (the balance between skills and challenges and the sense of control) are seen as important conditions for Mihaly Csikszentmihalyi's concept of flow[65]; an intensely demanding situation that elevates an individual to a level of optimal experience. The fantasy element can be likened to Csikszentmihalyi's mention of sports and other forms of play as conducive to flow because they create a space that is separate from

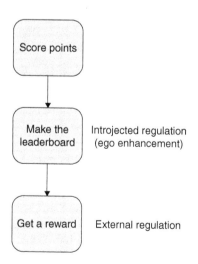

Figure 3.3   Motivation associated with the leaderboard mechanic

everyday reality; on the football field, other rules apply (this aspect of 'play' being separate from everyday reality will be covered more extensively in the next chapter). Curiosity into the workings of the game system and ways to cheat it are also mentioned as aspects that are stimulated socially in online games[66] (cheating will be discussed in Chapter 5). The issue of control becomes apparent when looking at human–computer interaction from a theatrical perspective, as Brenda Laurel[67] has done. One of the design principles she derives from theater is the first-person imperative: making sure that the interface represents what the person is doing instead of what the computer is doing. She states that the experience becomes much more enjoyable when the user participates as an agent, instead of just an observer. This richer, first-person experience can be seen in video games. It adds to the sense of control that was mentioned by Lepper and Malone. The sense of control is also heightened by the immediate feedback that players get in video games from both the game system and other participants, in the case of online games.[68]

In their Self-Determination Theory, Ryan and Deci identify a number of factors that can move extrinsically motivated behavior up the scale to more internalized behavior. In other words, they identify factors that increase intrinsic motivation. In relating their theory to video games, three motivating factors are determined[69]:

- experiencing autonomy (a sense of volition);
- competence (a feeling of effectiveness);
- relatedness.

Incidentally, these are also the three elements of motivation that Daniel Pink highlights in his popular book 'Drive',[70] basing himself on the work of Edward Deci, among others (but not relating motivation to video games). He labels them slightly differently: autonomy, mastery, and purpose. Autonomy and competence/mastery are closely related to concepts discussed earlier in this section. Autonomy for example relates to curiosity and discovery. The learning experience of mastering a game is also mentioned by game designer Raph Koster as a prominent source of the satisfaction and fun that video games can provide.[71] Relatedness is the feeling of belonging and being connected with others. The fact that the social aspects of online games promote relatedness was supported in a number of studies.[72] Relatedness/purpose is also closely linked to the concept of 'epic meaning' that game designer Jane McGonigal describes; the feeling of being part of something bigger than yourself that a large-scale online video game can give you.[73] The three most important motivating factors of video games are summarized in Table 3.1.

Returning to the subject of 'gamification', I hope to have made it clear that a simple points-leaderboard-reward game mechanic does not tap into these deeper sources of intrinsic motivation that video games can provide. And even the more sophisticated gamification projects may have their limitations. A recent study by Ethan Mollick and Nancy Rothbard of the Wharton School of the University of Pennsylvania[74] showed that even a well-designed gamification layer – that is, going beyond the simple leaderboard mechanic – did not increase intrinsic motivation in the work itself. The fun experienced in playing the game was seen as separate from the work. Mollick and Rothbard also touch upon the fundamental feasibility of using game elements and their corresponding play behavior in a business context because of the need for play to be free and voluntary. I will return to that discussion in the next chapter.

There is a final piece of criticism that can be directed at gamification, which is that it fails to acknowledge the complexity of games and play. Gamification is typically an approach based on general prescriptions, which its proponents say work in similar ways in many contexts. As in many other areas of business software, there is a focus on best practices and generic platforms. The starting point often is a collection of isolated game elements, not the player experience with its lack of predictability. These isolated game elements are taken out

Table 3.1   Motivating factors of video games

Motivating factor	Description
Autonomy	A sense of control or volition, stimulating exploration and curiosity, experiencing immediate feedback
Competence	Mastery, a feeling of effectiveness, a balance between skills and challenges
Relatedness	Purpose, the feeling of belonging and being part of something bigger than yourself

of the game context and are placed in another, non-game context. Because games are highly emergent systems, the effects that these elements have on player behavior in a different context are unpredictable and can only be known after the system has been put in motion. This means there is a potential for disappointment or even damage when a business tries to implement an off-the-shelf gamification solution or copies an approach that was successful elsewhere. In Chapter 6 I will expand on the process that video game designers have established to deal with this complex, systemic nature of games.

One perspective on the interest in gamification by the business community could be that it seems to indicate that the appeal of seeing the organization as a machine to which general prescriptions can be applied is still very strong. But as I mentioned in the previous chapter, the machine age in which Frederick Winslow Taylor's scientific management could be applied is far behind us. Yet the attempts to employ standard solutions or 'best practices' to complex problems seem hard to root out. Video games run the risk of becoming the latest victim of this tendency, because unsophisticated gamification projects are likely to lead to disappointment in the long run. And this disappointment could lead the business community to conclude that video games have nothing to offer for enriching organizations, which would be a shame because it is exactly this move beyond the machine age in which games can have an important role to play, as I will discuss in the course of this book. But for this role to be played out, it is necessary to embrace the complexity of games instead of reducing and shrink-wrapping it, which seems to be what gamification is doing.

3.4   What about simulation games?

This book explores the connection between games and organizations. So far I have limited the discussion to video games. But in order to be complete we need to acknowledge the thriving practice and academic tradition of simulation games as well. Simulation games are usually traced back to war games, simulated battles to test strategy and tactics. These war games go as far back as the seventeenth century, but reached new levels of sophistication during and immediately after World War II.[75] It was in the late 1950s that war games crossed over to business, with think tanks such as the RAND Corporation playing a pivotal role.[76] Over the years, some strands of simulation gaming such as Microworld simulations[77] have tended more towards an analytical, mathematical approach. Others have stressed the use of simulation gaming as an instrument of organizational change.[78] Combining those two perspectives is the use of simulation gaming for strategy development, which opens up the design process by using participative modeling.[79] Overlooking the past four decades of their use, there are five major goals that can be associated with simulation games[80]:

- achieving a learning objective;
- improving decision-making skills;
- improving communication and teamwork;
- strategy formulation and policy making;
- gaining experience and insight about a future situation.

An important element that unites all these different applications of simulation gaming is the process of experiential learning. David Kolb's Experiential Learning Theory[81] sees learning as a cycle consisting of four modes: feeling, reflection, thinking and action. Elements such as role-play and an organized debriefing help make simulation games a safe environment for participants to pass through this cycle, sometimes more than once.

Most managers working today will have been involved in a simulation game of some sort in either a training situation or an organizational change project. Practitioners and academics have jointly turned simulation gaming into a mature field with its own professional association, academic conferences, and journal. But even though simulation games and video games share a name and both came of age in the latter part of the twentieth century, the two fields have developed almost entirely separately. Only recently, as video games have become a major cultural phenomenon and video-game scholarship has established itself as a credible branch of inquiry, has the field of simulation gaming hesitantly begun to direct its attention in that direction.

The simulation gaming tradition shows that using games in a business context is nothing new. However, the rise of video games merits a renewed investigation of this intersection. With the growth of the video game market, so has the number of video-game designers, video-game scholars, and the sophistication of their fields grown. New insights have been unearthed and new instruments have been developed. For example, video-game designers and game scholars point to the importance of emergence, in playing a game but also in designing it. This emergent quality of games and game design and its relation to complex systems provides a new perspective on the use of games as a tool for business. This is among the aspects we will explore in the forthcoming chapters.

4
Play

On a pleasant autumn evening in 2008, 15 people got together in a hotel meeting room in the Dutch seaside resort of Bergen. All of them had some involvement in setting up a new Elective Care Center (ECC) as part of a nearby hospital. Most of the participants in the meeting did not know any of the others. They had not received an agenda, just a one-page invitation with a photo of a girl blowing bubbles and the request to join an important workshop in the design of the new center. No preparation was necessary. The people trickled in, had some coffee, and stood about somewhat uneasily. Some talked about the nice weather or the trouble they had in finding the hotel. Among the participants were a surgeon, a general practitioner, a nurse, a health insurance representative, and even a patient. Although the surgeon and the general practitioner had never met, they knew each other by name. The surgeon decided to take this opportunity to complain about some of the wrong referrals he had received from the general practitioner. The general practitioner responded with his complaints about the delays in receiving the results of procedures. The mood was set for a traditional round of exchanging viewpoints without making much progress towards the shared goal of a new center. But this was not going to be a traditional meeting. In the course of the next four hours, participants would use posters to make caricatures of the group they represented, compete for who could write the most activities on small white cards, and attempt to lead the group in a brainstorm about how they could best achieve their goals. By the end, the surgeon and the general practitioner were exchanging business cards and resolving to stay in touch and improve their collaboration. It was an example of the power of a concept that is very basic to human culture, but often far removed from business environments. It was the power of play.

4.1   Games and play

Since the previous chapter dealt with games, let me try to untangle play from games. The notion of play usually evokes a mental image of laughter, movement, and the outdoors. Play is an activity, a type of behavior that almost everyone has an intuitive understanding of. And a game is one of the ways to evoke that behavior. But however intuitive the notions of play and game are, they remain difficult to capture in a definition.

   At the beginning of the twentieth century, the Dutch historian Johan Huizinga was one of the first to attempt to extensively characterize play. Huizinga explored the play element of culture in his book *Homo Ludens*.[1] In his opening chapter, he set out to describe the main characteristics of play and identified a number of elements that have become important philosophical underpinnings of the video game design discipline and the academic field of game studies. The first and most important characteristic of play according to Huizinga is that it is a free (in the sense of voluntary) activity. You cannot force someone to play. According to Huizinga, ordered play is not play. Immediately, we can see a formidable barrier forming against introducing play in the context of work or business. That is a theme we will return to several times in this book. I am reminded of the first time I presented a paper about the use of games in a business context at a conference for video-game scholars in the spring of 2007. As a management consultant associated with a business school, I was already the odd man out. But after I presented my utilitarian view of games – admittedly not very sophisticated as I look back on it today – I was met with a barrage of questions, phrased constructively but nonetheless with a skeptical undertone. 'You cannot force people to have fun,' was one of the comments that stood out for me. The premise being, apparently, that the corporate world was about forcing people to do things. The discussion was much more nuanced than that, to tell the truth, but it was obvious that I had made the beginner's mistake of not taking into account Huizinga's 'first law of play': it is free and voluntary.

The recent paper by Ethan Mollick and Nancy Rothbard that I mentioned in the previous chapter in fact shows empirical evidence in support of Huizinga's first law. In their study, Mollick and Rothbard assigned salespeople to one of three conditions. The first was a game condition, in which salespeople were presented with a game layer (designed by professional game designers) on top of their usual reports of sales figures. There was a control condition in which no changes were made to the work environment. There was also a leaderboard condition, in which a scoreboard presented data the salespeople were already familiar with, but in a more attractive manner. What they found was that the game condition did not affect performance in a positive manner, but that it did have a positive impact on feelings in the workplace. But this positive impact only occurred if employees had consented to the game. Consent was operationalized here as following the game, understanding the rules, and considering them to be fair. If there was no consent, there was even a negative impact on performance. This same negative impact on performance was present in the leaderboard condition. So as I mentioned in the previous chapter, this important study (one of the first empirical investigations of gamification in the workplace) contains an important warning for superficial, leaderboard-type gamification projects. They may harm performance. With regard to consent, the study confirms Huizinga's basic tenet that ordered play is not play, at least not in the sense that it evokes positive emotions. Without referring to Huizinga, Mollick and Rothbard call this 'the paradox of mandatory fun' in their paper. A final interesting result from their study that is worth mentioning here is that an important predictor of consent was found to be the exposure to multiplayer video games outside the work environment.

Apart from the aspect of voluntariness, Huizinga goes on to talk about another crucial characteristic of play. He states that play is not real life. It is a departure from the ordinary into a temporary sphere of activity. It 'plays out' within certain limits of time and place. Here Huizinga touches upon another elemental quality of play. We can enter into and step out of this temporary world. Inside, we may act tougher or be less inhibited; we may dress differently, or we may physically assault people with whom we will later have a drink. The notion of this sphere in which play transpires should not be taken too literally. It is a permeable boundary.[2] Of course social relations and the culture at large do not disappear when play commences, but new meaning does arise. And this new meaning arises out of another characteristic that Huizinga identified, which is the role of rules. With the concept of rules, we enter into an area of overlap between play and game. In fact, when referring to rules, Huizinga talks of games. He says that every game has its rules.[3] The rules give rise to new meaning. Suddenly I am not allowed to touch the ball with my hands. Or I have to wait for my turn. Or I can only use gestures to get a sentence across to my friend. The rules circumscribe the temporary space of the game. I use the word 'game' here because I would contend that a game is always rule-based, whereas 'play' is a broader concept that can be free of rules.[4] This distinction leads us to Caillois.

In the second part of the twentieth century, the French philosopher Roger Caillois wrote *Les jeux et les hommes*,[5] in which he contests some of the elements of Huizinga's description of play. In essence, Caillois agrees with Huizinga on the two most important characteristics of play: that it is free and voluntary, and that it is carefully isolated from the rest of life. But Caillois does not see rules as a precondition for games. He states that creating a fiction is another way to bring about this game space, this separation from reality. Caillois sees a continuum from fiction-based, improvised games to rule-based, organized ones. He terms the improvised side of the scale 'paidia', after the Greek word for free-form play. The organized side he calls 'ludus', after the Latin term that denotes both game and play. Caillois sees several categories of games that all contain this continuum. In this classification, he sees games as being much broader than Huizinga does, including also games of chance and even theater. It is not useful for our discussion to go into Caillois's extensive classification here. What we can learn from Roger Caillois is that there are two essential characteristics of play (which he and Huizinga agree on) and that rules are what distinguish organized games from free-form play. The former has them, the latter not necessarily so.

Let me introduce a third thinker to shed some more light on this isolated space in which play takes place. Whether or not we agree with Caillois that a fiction can create that space as effectively as rules can, it is clear that a particular attitude is needed to enter into and maintain the play space. And since play is voluntary, players need to be invited to assume that attitude. They can always decide that they don't want to play anymore, albeit running the risk of being called a spoilsport by the rest of the group (more about spoilsports in the next chapter). The philosopher Bernard Suits has called this particular attitude when entering into the play space the lusory attitude (from the Latin 'ludus', for game and play). He points out the paradox inherent in games, which is that certain efficient means to attain the game's goals are prohibited by the rules. For example, in golf it would be much more efficient to take the ball, get in the cart, drive to the hole, and deposit it with your hands. There is no rational reason not to do this. It is not dangerous or harmful to anyone. Yet golf players see this course of action as impossible, because they have assumed the lusory attitude of the game. As Suits phrases it, they have voluntarily accepted unnecessary obstacles.[6] Although Suits limits his definition to games, we could extend it to the free-form (paidia, in Caillois's terms) end of the play spectrum. If children play with their Lego, they choose to maintain a certain fiction. This airplane cannot really fly, but let's act as if it can. Of course for children, assuming this lusory attitude is not a deliberate choice. It has much more to do with how they see the world and with play being an instinctive way to learn new things in a safe environment. That is not a perspective we will explore here,[7] but this familiar image of children playing clearly contributes to how we as adults look at the play activity, at least at the free-form paidia side of the spectrum. It can be associated with immature behavior.

At the rule-based, ludus side of play, Bernard Suits's insightful deconstruction of the concept points us to an important barrier if we want to introduce games or game elements in a business setting. Why would we create unnecessary obstacles in an organization? One of the more popular management philosophies of the past decade has been Lean Six Sigma,[8] which concerns itself with improving business processes by eliminating 'waste'. That seems to prohibit imposing unnecessary obstacles and thus blocks the introduction of game elements. And yet, it is not that black and white. The Toyota Production System[9] that lies at the base of the Lean philosophy stresses the importance of experimentation in the context of problem-solving. And experimentation or trial-and-error behavior in a safe environment could be considered an integral part of play. It is one of the things that make play such a powerful tool for learning. It seems there are different ways to look at play from a business perspective. I would say that many managers still dismiss the concept in itself as something for children. But at the same time, a company such as Google is often praised or even envied for being playful. Exemplified by offices full of slides, game tables and free food, and by a relative autonomy given to its employees, this playfulness is credited for generating and sustaining the creativity and innovation that has made Google the 46 billion dollar business[10] that it is today.[11] As I said in Chapter 2, this is probably tied to the specific context of Google. But there is definitely a link to be made between play and innovation as there is between play and idleness. This paradox is what play theorist Brian Sutton-Smith has called 'the ambiguity of play'.[12] Sutton-Smith tries to bring some coherence to the fragmented discourse surrounding the subject by identifying seven 'rhetorics of play', seven ways in which certain groups frame play in accordance with their own beliefs. This helps in understanding the different ways in which the business world views play. The view that still seems to dominate, especially in this time of economic crisis and hesitant recovery, is what Sutton-Smith calls 'the rhetoric of play as frivolous'. It is the view that uses the label 'play' to denote 'activities of the idle or the foolish'[13] that have no place in a serious work environment. It means seeing play as immature behavior, as something for children. But another perspective on play is gaining ground in the business world. It is the 'rhetoric of play as the imaginary' combined with the 'rhetoric of play as progress'. The former 'is sustained by modern positive attitudes toward creativity and innovation'[14] and is closely aligned with concepts such as improvisation, which have also gained a foothold in management theory.[15] It is the frame that is often used for talking about the playfulness that seems to characterize Google. It is considered a positive characteristic of that organization, because it is seen as having a connection to Google's capacity to innovate, create new products, and branch out into new markets. Again, what the mechanism looks like that connects playfulness to business performance in the case of Google has not been convincingly demonstrated, to my knowledge. But the fact that Google is such an iconic company in our contemporary business landscape makes these rather superficial attributes an attractive template, with all the associated risks discussed in Chapter 2. The rhetoric of play as progress is another positive lens that is increasingly applied to play in a business context. The view that players of MMOGs such as *World of Warcraft* (discussed in the previous chapter) are learning new skills that could be of use in an organization fits with this rhetoric, as does the use of simulation gaming (also discussed in the previous chapter) to have organizational actors go through a process of experiential learning. Play as progress sees play as a safe environment for learning. But when the learning environment and the target environment (in which the skills need to be put to use) are distinct, this perspective carries with it the persistent question of how to transfer what was learned from one environment to the other. Unfortunately, when applying video games, serious games, or simulation games as tools for learning in a business environment, this question always needs to be addressed and often forms an obstacle to the effectiveness of these instruments.[16]

There are signs that organization and management scholars have lately become more interested in organizational play, in 'play as progress'. Ronit Kark has explored the use of play to develop managers' leadership skills.[17] She posits that play will have a positive effect on leadership development by creating a bounded, psychologically safe space, shielded off from the pressures of social validation that can exist for organizational leaders in their everyday work environment. She also foresees positive effects of identity play in claiming and granting leadership roles, and of using play for experimentation with intrapersonal and interpersonal leadership skills such as decision-making and emotional intelligence. Kark is not talking about video games, but we can relate her propositions to some of the leadership phenomena witnessed in *World of Warcraft* (described in the previous chapter).

David and Alice Kolb have studied how play can create a learning space, specifically one aimed at promoting the concept of experiential learning.[18] Using a pick-up softball league as a case study, they show how deep learning takes place in this 'ludic learning space'. Two important qualities of this softball league case study may distinguish it from other learning spaces. One is the fact that the informality of the league allowed players to create their own rules to a certain extent. The other is that this ludic learning space was not based on one session, but was replicated each week as a group of players would show up for their regular game of softball. But more rule-bound, single-session learning spaces such as those created in simulation gaming (discussed in the previous chapter) have also been shown to engage the experiential learning cycle of experiencing, reflecting, thinking, and acting.[19]

Charalampos Mainemelis and Sarah Ronson see play as 'the cradle of creativity in organizations'.[20] They stress the capacity of play to produce novelty and to experiment with alternate realities and they propose the creation of play spaces, as temporary breaks from the organization's normative structure. Mainemelis and Ronson suggest that what is generated in these play spaces (such as new ideas for products) can be taken back into the organizational reality and that these play spaces help in building a disposition in organizational actors and a social context in the organization that is conducive to creativity.

I believe that some of these authors on organizational play may attach too much value to the magic of the play space. Largely on a theoretical basis, they conceive of many forms of learning and creativity taking place there. At the same time, the field of game studies has begun to question the firmness of this boundary between play and everyday reality. As I said earlier in this chapter, it is certainly true that new meanings arise when the rules of a game are set in motion, but existing social relations will not completely disappear (unless of course basic attributes of players are masked by their character on the screen, as in *World of Warcraft*). At the very least it is important to acknowledge that setting up this play space inside an organizational context is a difficult and delicate undertaking. The discussion of play in this chapter suggests that using it effectively will require creating an isolated and temporary context that organizational actors need to be carefully invited into. I will call this context a 'lusory space'. And although I am not as optimistic as some of the organizational scholars mentioned above, I agree with the basic premise that play can encourage creativity and can lead to learning experiences. But what I would like to focus on is not so much the wonderful things that will happen once we enter this lusory space, but rather how we would go about creating one that is effective.

4.2   Lusory spaces in practice: The Elective Care Center

In this section I will give an example of a project in which lusory spaces were created in an organizational context, followed by a number of practical recommendations.

4.2.1   Setting up an Elective Care Center

In the spring of 2008, one of the largest non-university hospitals in the Netherlands established a new organizational unit, the ECC.

This new center was to be a separate entity on a different location, with the aim of providing low-complexity care that can be planned well. This type of care included procedures such a hip and knee surgery and cataract operations. The center was set up to compete with similar services by private clinics, in the newly liberated Dutch healthcare market. To be competitive the hospital wanted to emphasize innovation and new ways of working and organizing. An important element of this was a different attitude towards patients: the patients had to be approached individually and as an equal partner. If desired, they should be able to direct part of the care process themselves. Discussions about setting up a separate ECC had been going on for a long time at the hospital but the organizational design of this new center had finally started in early 2008. Within that context, I was asked together with a co-researcher to direct a project in which the points of departure would be defined for the ECC. These points of departure should then be used to further develop the five design variables that had been identified for the ECC organization; care process, real estate, ICT, human resources management (HRM), and general management. The request to consult on this project was posed to me by the managing director of the ECC. This is relevant, because he is someone who deliberately seeks out innovative ways to solve organizational problems and who is not afraid to enter into a process of which the outcome is difficult to determine beforehand. A game-based approach piqued his interest and a brief explanation of the underlying logic was enough for him to give the go-ahead.

The introduction of a partly free market in the healthcare industry in the Netherlands was an important reason for why the ECC came into being. At the same time, the new healthcare services that sprung up as a result of the free market were looked down upon by some doctors. This negative image of private clinics formed an obstacle to setting up the ECC. It mostly had to do with the status of the physicians providing this type of care. A surgeon who does knee operations all day just does not have the same standing among his peers as one who performs open heart surgery. But those involved in setting up the ECC understood the necessity to separate complex care from elective care. They felt it was necessary to organize elective care in such a way that it comes close to the atmosphere of a private clinic. If they wouldn't do that, then they were convinced they would lose out to the centers that do give that attention and are able to offer efficiency and patient-friendliness.

The different type of care that would be provided in the ECC would mean a different way of working for both doctors and nurses. This would appeal to some and not at all to others. One of the important characteristics of care in the new ECC would be that there is much more of a patient-focus in comparison with the way of working at the hospital. The current situation would have to be partly reversed. In the future, the doctor would come to the patient, sometimes literally. And that new, patient-focused attitude would require flipping a switch, most of all for the doctor. One of the most important problems for the ECC was the resistance by these physicians. However, some of those involved in the ECC felt that the attitude of physicians was already changing and that it was possible to get them enthusiastic about new ideas.

4.2.2   Setting up lusory spaces

To help the hospital establish the points of departure for their ECC, we proposed an approach that would result in a 'meta-design': a set of rules from which the ensuing design variables (care process, real estate, ICT, HRM and general management) could be developed further. This approach was based on the video game design process, which is something we will cover later on in this book, in Chapter 6. What is relevant for the current discussion is that the approach is based on a series of workshops (see Figure 4.1). The goal of the workshops was to incrementally produce the meta-design. Each workshop individually was intended to create a lusory space, in order to make them as productive as possible and to create energy with participants that would extend beyond the boundaries of the workshop.

To create these lusory spaces, a number of techniques were used First of all, the workshops were tightly scripted, almost down to the minute. These scripts were strictly adhered to and enforced by the moderators, who could be described as game masters. The moderators took a somewhat detached position and never got involved in the content of the discussions taking place between the participants. The moderators were solely responsible for setting up the framework for the brainstorms and discussions. What happened inside that framework was the responsibility of the participants. The detached or even stern attitude that the moderators took was compatible with

Figure 4.1   A scene from one of the workshops at the ECC

their enforcing the rules, which were never questioned. Those rules were concerned with the time allotted for a certain activity, who was allowed to do what, and the conditions for winning. The moderators frequently reminded participants of the time that was left and they determined the winners. Competition was an essential part of the workshops. Most activities were competitive in some way. Usually, the reward for winning was not made explicit, and often it was nothing more than the opportunity to structure or adjust the results of a discussion.

Another technique that was used to create a lusory space was supplying only limited information beforehand about the workshops. Instead of a meeting agenda, there was just a one-paragraph description. Also, ambiguous or somewhat odd assignments were given. For example, the participants of the first workshop – a core team of three people – received a homework assignment that asked them to invite additional participants for workshop 2. But they were not allowed to use certain words in their invitation.

To a certain extent, role-playing was also used to create the lusory space. In workshop 2, we had gathered together the most important players involved in setting up the ECC, such as a surgeon, a general practitioner, a nurse, a health insurance representative, and a patient. For each of these players, a poster had been designed with an illustration that represented this group. Participants were asked to give life to this character. What was his or her name? How would you describe their character traits? What is their motto? Most participants did not need much encouragement to take some markers and start filling in the blanks of their virtual representative. The surgeon in particular made a considerable effort in creating a caricature of his profession, a stubborn bone sawyer who only cared about having the next anaesthetized patient on his operating table on time. This created room for the others to discuss the surgeon's way of working, without it feeling like personal criticism. Some participants took the identification with their character so seriously that they only responded to the character's name and no longer to their own for the rest of the workshop. But those were the exceptions. The differences between the participants in their inclination to perform (in the sense of acting out a role) clearly showed in this exercise, but there was a general atmosphere of participation and fun.

A final technique, and perhaps the most obvious one, to create a lusory space is to play an actual game. This was done in the final workshop. In the coming chapters, I will talk more about the purpose of such a game and how it is produced. Strictly in terms of physical appearance, it is a board game, albeit one that is custom-made for this specific project and whose content reflects the organization in question. In this case, the aim of the game reflected the goal of the ECC: to prevent and cure as many non-complex ailments as possible. In the game this goal could be attained by building the ideal ECC. However, time and budget were limited and collaboration with other players was necessary. Playing a board game is an activity familiar to most people, albeit not in a business context. There always seems to be a number of phases when using a board game to create a lusory space. At first there is some reservation among participants as they get used to the fact that they are going to play a game with these people, most of whom they do not know. They listen to an explanation of the rules, which are inevitably complex when put into words. Then players start playing, going through the different steps prescribed by the rules, and starting to grasp the mechanics of the game.

It is then that the lusory space comes into existence and the energy level begins to rise. Players start figuring out how they can win, either collectively or individually, and a certain zeal becomes visible. Speed picks up, laughter is heard, and players commit to what is going on at this table, sometimes without much regard for what their functional or hierarchical relationships are supposed to be with the others. The lusory space is in effect and the playing field has been leveled.

4.2.3   Recommendations for setting up a lusory space

Based on the ECC project and others, I can formulate a number of recommendations for setting up a lusory space in an organizational context.

4.2.3.1   Create competition

The competitive element was an important factor in the workshops. Some participants clearly showed their eagerness to be the winner of an assignment. Winning in most cases meant generating the most ideas, which brought the aim of the game in line with the aim of the workshop. The prospect of a reward is not a prerequisite for creating competition. Being the winner is usually enough of a reward in itself, although there are clear differences in competitiveness between people. Competing against the clock was another element that caused emotion. The moderators regularly mentioned the remaining time for an assignment. One of the things that time pressure does is prevent extensive discussions from taking place. Decisions and choices have to be made rapidly, which can be a welcome departure from regular meetings.

4.2.3.2   Omit information

The members of the core team of the ECC expressed their uncertainty during the first two workshops about where the process was headed. Afterwards, they said this lack of information fitted with the game. It caused excitement. After each workshop, they were curious what the next step would be. Some of the other workshop participants felt more uneasy about the lack of information they received beforehand.

> We were kept guessing because we didn't receive a meeting agenda. And when I checked who the other participants would be, I hardly knew anybody. I thought the invitation could perhaps have been handled differently.

Even though it caused uneasiness in some, this way of inviting people was an essential part of creating the lusory space. The core team of the ECC was given strict instructions by the moderators to give only very limited information when inviting participants. But some of the people that they spoke to wanted to know more. They wanted to know who they would be representing or what business they would be doing with the hospital later on. The core team quickly decided that these were not the people they wanted in the workshops. They looked for people who they felt would contribute freely and that were willing to 'enter into this adventure with them'. Because of the lack of information, participants entered the workshop fairly unprepared and their motivations for participating centered on curiosity and meeting new people. A self-selection had taken place of participants likely to be open to entering a lusory space.

4.2.3.3 *Don't let the game take over*

There was an important tension that played a role during the workshops between the game elements and the content of what was being discussed. During the final workshop, when the goal of the game was to build the best possible ECC, it seemed that the game element sometimes took precedence. Cards were played not because of valid arguments but because the rules of the game allowed it ('Do we have enough money? Okay, let's build it'). One participant later talked about this delicate balance between game elements and content, which has to be monitored constantly.

> The trick is not to make the invigorating game element too much of a distraction for the content, but the reverse is likewise true. You see that people don't worry anymore about whether their contribution makes sense or what their colleagues will think of it. They are committed to what happens at that table. Yet the game element is not so dominant that people have the feeling that what they're doing is silly.

4.2.3.4 *Enforce the rules*

What characterized the organization of the workshops was a strict adherence to the rules. The attitude of the moderators that 'rules are rules' was immediately accepted by the participants without much comment.

> So then I had to write something on all these cards and I thought: this is absurd, but I'll do it anyway.

Clarification was asked regularly about the rules, but they were never questioned. The competitive element did cause some participants to look for the boundaries of the rules if that could help them be the winner of a round. The rules are in essence what created the lusory space. Creating and maintaining this space is a delicate undertaking. There were moments when the participants of the workshops entered into and stepped out of the lusory space. The latter occurred for instance when the participants themselves had to lead a brainstorm – and thus enforce the rules – in one of the workshops. The number of ideas generated during this brainstorm was fairly limited. It became a discussion rather than a game, which drained the vitality from it. Participants started to drift off and check their phones. The lusory space was broken. On the other hand, when it was intact the organizational context and hospital hierarchy seemed to fade away.

Rules are an essential part of creating a lusory space. But there are many other dimensions to rules in an organizational context, as I will discuss in the next chapter.

5
Rules

At the offices of We Beat The Mountain (WBTM) – a start-up company that designs, produces and sells products from recycled materials – we had entered the last stage of our meeting. As WBTM was about to launch its first consumer product, the founder and his core team were assembled for this workshop, which used lusory spaces to brainstorm about aspects of their organization such as the relevant stakeholders and the areas of knowledge that were essential for WBTM. Together with my fellow moderator I had just explained what the objective of the last round was: to write down on the colored cards that were in front of them, different things that WBTM could spend its money on. The winner would be the one who wrote down the most items that others also wrote down. The intention was to have participants think about what they wrote down instead of going for pure volume of ideas (which had been the aim of a previous round). That was how it had been designed. But that was not how it went down. After I had started the timer for the five-minute round, the WBTM founder looked around the table at his core team and said, with a mischievous smile on his face: 'Is there anyone who wants to win this round?' No, actually there wasn't anyone who really wanted to win or at least no one admitted it. 'So then why don't we just do this exercise out loud and agree on what we write down on the cards, so everyone will be a winner?' One of the other team members looked at me. 'I don't think the game master would agree with that,' he said. But all I could do was admit that there was nothing in the rules that prohibited this behavior, so it was allowed. The founder had discovered a weak spot, a loophole in the rules. The brainstorm went ahead, but the lack of competition seemed to lower the quality of the ideas. It was a reminder that rules can have unintended consequences.

5.1   Being a spoilsport, cheating, and gaming the system

Rules govern conduct within a particular sphere. Sometimes they describe what is possible, but most of the time rules prescribe what is allowed. It is possible to touch this painting but it is not allowed. Rules can apply to grammar, to religious practice, to conduct in a museum, a school, or in the military, to how a society is organized (these rules are usually institutionalized in the form of laws), or to what is morally speaking right or wrong. Rules can be obvious, internalized, or almost intuitive to most ('thou shalt not kill') or they can be frustrating and illogical. Understanding the rationale behind a rule can help make it acceptable. If everyone were to touch the painting, then it would be damaged and no one would be able to enjoy it.

In the context of play and games, rules have a very specific function, as we saw in the previous chapter. The rules of a game create unnecessary obstacles – in the words of Bernard Suits, the philosopher discussed in Chapter 4 – to reaching a certain goal. That makes many of the rules in a game illogical. There is no clear rationale behind them outside the sphere of the game, but you are just going to have to follow them if you want to enter into this lusory space. Perhaps there are more efficient ways to move the knight, but then you would not be playing the game of chess. And that is what we agreed we were going to do. Playing a game demands that you abide by the rules. So what happened in the example I gave in the opening of this chapter?

If we turn back to Johan Huizinga – the historian whose work *Homo Ludens* I discussed in the previous chapter – we see that in the breaking of rules he distinguishes between the spoilsport and the cheat. The spoilsport is the one who does not acknowledge the rules of the game and acts as if they did not exist. He or she would catch a ball with their hands during a game of soccer, run to the opposing goal and throw it in to score. Unless the other players decide to accept and follow this behavior (and thus create a new game), the spoilsport is usually met with scorn. In organized sports they are neutralized by the referee. According to Huizinga, this is because spoilsports tear down the game world. They show the limits, the fragility or perhaps the absurdity of the game world and thus form a threat to the other players. On the scale introduced by Caillois (also covered in the previous chapter) from 'paidia', free-form play, to 'ludus', organized games, I would say that the spoilsport hardly forms a threat on the ludus side of the spectrum. At that far end of the scale, games have become so organized and institutionalized that the spoilsport does not stand a chance. Players have either become dependent on the game for their livelihood (a professional poker player), they respect the rich tradition a game has (a player of the Chinese game Go), or the game has become so entrenched in society (soccer in Europe, baseball in the United States) that being a spoilsport would amount to an act of cultural rebellion. The only instances in which we can witness the spoilsport in organized games are when unwritten rules are not acknowledged – for example, if a soccer team kicks the ball outside the lines when there is an injured player on the field and the opposing team does not return the ball to them when play resumes, or in American football, if the quarterback takes a knee at the end of a game to run the clock down and the opposing team nevertheless charges the offensive line instead of accepting that the outcome of the game has been decided. Willfully ignoring these unwritten rules is usually a sign of fierce rivalry between two teams and is generally regarded as unsportsmanlike conduct.

Video games form a special category, because inside a video game the computer upholds the rules and it is a ruthless arbiter. So it would seem that the only possible course of action for the spoilsport is to quit the game. Strictly speaking, this is true. But in multiplayer games – especially in the MMOG variety discussed in Chapter 3 – there is usually more than enough room inside the confines of the rules for what is called 'grief play'; deliberately obstructing, annoying, or generally showing your superiority over other players just for the fun of it. In spirit, we could certainly consider these players spoilsports. They are not able to threaten the game world as a whole, but they can sure spoil the fun for individual players.[1]

If we move to the paidia side of the spectrum, games become more fragile and more susceptible to spoilsports. The lusory spaces I described in Chapter 4 certainly are vulnerable in that sense. Because there is usually such a contrast between the average workday and the lusory space that you ask participants to step into, there is always the danger of them not accepting the rules, even though I've never personally encountered such a situation. The reaction to a lusory space described at the start of this chapter may seem like that of a spoilsport, but I would not call it that. I will get back to what I would call it in a moment. Even though an outright questioning of the rules of a lusory space is rare, I have suspected some participants of being passive aggressive spoilsports. What I mean is that participants in a workshop listen to an explanation of the rules and think to themselves: 'I don't think I want to play this game.' Unless the spoilsports think they can get the group on their side (which I have never seen happen), they have no choice but to pretend to go along with the rules. And it is likely they will get caught up in the energy of the game after a while, or will start appreciating the rules when they understand the gameplay that these rules set in motion. One thing that helps in neutralizing spoilsports is to be very clear and strict when explaining the rules, to act as if they are the most logical thing in the world. There will be no discussion about them and there is no room for amendments. This confident attitude of the game master (something I also mentioned in the context of lusory spaces in the previous chapter) is only possible if he or she has tested the rules and knows what the consequences will be of setting them in motion. This is a subject we will return to later in this book.

In addition to the spoilsport, Huizinga also talks about the cheat. The cheat is a different animal, and far more prevalent. Cheat pretend to respect and follow the rules, but stealthily behave in ways that are not in accordance with those rules in order to gain an advantage over the other players or to get to the end of the game quicker (in the case of a single-player game). In the sense of preserving the game sphere, cheats are less dangerous than spoilsports, because they want to play the game. It is just that winning is so important to them or they are so bad at the game (as prescribed by the rules) that they look for ways to gain an advantage that can be considered unfair towards those who are following the rules. Spoilsports usually do not care about getting caught. Their acts are out in the open. But cheaters try to conceal their acts as long as possible. In organized sports, the officials are the ones charged with detecting the cheat and handing out penalties. This can range from a free kick after a rough tackle in soccer to a ban from the sport altogether such as after the use of banned performance-enhancing drugs in cycling. The latter is an interesting example because in cycling there seems to have existed a time when almost everyone was cheating. The Dutch cyclist Michael Boogerd – who confessed to doping a few months after Lance Armstrong – motivated his cheating by saying he was fed up with being passed left and right by those who were using the banned substances.[2] The way he described it, everyone was cheating and he was left as the fool who hadn't understood the rules had changed. If everyone is doing it, is it still cheating? Without wanting to enter into a discussion of morality, this example does point to the fact that cheating is sometimes hard to pin down. If I grab my leg and grimace even if I don't feel any pain after a tackle in soccer, am I cheating? Is using a walkthrough from YouTube that shows me how to finish a level in Tomb Raider cheating? Is counting cards during blackjack cheating? Was having the repairman set Space Panic to unlimited lives cheating? Was what the founder did during the WBTM workshop cheating?

In fact, I would consider what he did was neither being a spoilsport nor a cheater, if we follow Huizinga's description. He respected the rules and was very open about his intentions. All he did was point out an alternative way to follow the rules. He demonstrated the unintended consequences of the rules and showed that a change to the rules would be necessary if this alternative behavior was undesirable.

Unintended consequences often plague rule systems. The 2008 financial crisis put the spotlight on a collection of rules about how compensation was determined, that turned out to be perverse incentives. That is to say, these incentives had consequences that were undesirable, at least when viewed from the perspective of the economic system as a whole. These incentives included lenders being paid full commission upon closing a mortgage, and thus lacking a motivation to care for its long-term viability.[3] There was and still is the fact that ratings agencies are paid by the issuers of securities and not by the buyers (the latter being the ones bearing the risk).[4] And where compensation of executives is based on a company's financial results, either directly or indirectly (through the share price), we have seen many examples of manipulating those results, which led to the downfall of companies such as Enron[5] and WorldCom.[6] But these perverse incentives are not limited to the financial industry. In the Netherlands as well as in many other countries, doctors are compensated for treating patients and not for preventing medical conditions.

Treating more patients thus means more income for doctors, which is one of the sources of the rising cost of healthcare and is detrimental to the system as a whole.

Most of these examples involve a certain variable being used as a proxy for performance, on which compensation is then based. Most human actors will then attempt to maximize the variable being measured in order to maximize their compensation. Mind you, they are not cheating. They are playing by the rules. But they may very well be aware that they are acting contrary to the spirit of the rules. That is, they are likely aware of the goals for which the rules were put in place and realize that their actions are in conflict with those goals. But as long as they stay inside the confines of the rules, they have no qualms about letting their personal goals prevail. This behavior is sometimes called 'gaming the system', although I would argue that this phrase carries with it a somewhat negative connotation of malice. The surgeon maximizing the number of procedures he can do on one day would likely not consider himself to be gaming the system. He is just maximizing his income within the confines of the rules, as any *homo economicus* would. The line becomes a bit more blurred when we look at the example of tax evasion. If Google, Amazon, Starbucks, or Apple ensure they pay a minimum amount of taxes by letting revenues flow to the most advantageous tax climate, even if their economic activity in that locale is virtually nonexistent, are they gaming the system? It is clear they are following the rules, as their representatives have stressed to parliamentary commissions in both the US[7] and the UK.[8] And we can argue about whether they are going against the goals of the global economic system. Fewer taxes paid mean more income available for investment or for shareholder dividends. But judging by the attention paid to this tax evasion both by the media and by politicians, there is something about it that doesn't feel right. As Margaret Hodge, Chair of the UK Public Accounts Committee said to a Google executive during a hearing: 'We're not accusing you of being illegal, we're accusing you of being immoral.'[9] If we look at it on a national level, it is easy to see what is wrong. If a government has a company inside its borders with considerable economic activity – let's say a chain of coffee shops – but according to its books there are no profits that it can levy taxes on, it can feel cheated (even though strictly speaking it has not been). Whereas a neighboring country with a lower tax rate may be the

Table 5.1 Being a spoilsport, cheating, and gaming the system

Spoilsport	Cheat	Person gaming the system
Does not acknowledge the rules	Acknowledges the rules but does not follow all of them	Follows the rules
Does not support the goals of the system	Supports the goals of the system	May or may not support the goals of the system
Knows he or she can no longer play the game	Wants to keep playing the game	Wants to keep playing the game
Is outcast by the other players	Is penalized and possibly outcast, if detected	May be admired or denounced, depending on your moral stance, but cannot be outcast or penalized
Is a threat to the system	May become a threat to the system if many start mimicking this behavior	Highlights a design flaw, which may lead to improvement of the system

beneficiary of tax income from economic activity not taking place there. Is the neighboring country gaming the system or is it the multinational corporation? In any case, if the major forces in the global economic system were to agree that this type of behavior is undesirable, they would adapt the rules. In that sense, gaming the system serves a purpose. It can show that the rules that govern a system contain a design flaw. I would even go as far as to say that the person gaming the system is not doing anything wrong. The responsibility for the undesirable outcome lies with the designer of the rules. A game designer actually welcomes players who attempt to game the system, who explore the boundaries of the rules. They are an invaluable help in rooting out the problems that a rule set may have. This is something we will talk about more in Chapter 6. To conclude this section, I will summarize the attributes of the spoilsport, the cheat, and the person gaming the system in Table 5.1.

5.2 Simple rules and complex patterns

The examples of gaming the system in the previous section show that rules can be very dangerous if they are not wed to a clear understanding of the behavior they evoke. Of course, if rules would simply prescribe or prohibit behavior ('don't feed the animals') it would be very simple. But a rule never exists in isolation. It always interacts with other rules. Even a few very simple rules can create a complex and unpredictable range of possibilities. This becomes apparent if we attempt to analyze this range.

Game theory is a method used in economics as well as other disciplines for modeling the interactions between rational decision makers.[10] Despite its name, game theory has very little bearing on games of the type discussed in the previous two chapters. What game theory does is chart a possibility space based on moves that are available to the players, combined with the possible pay-offs of these moves. Available moves could be considered a narrow form of rules. Normally, the rules of a game give a general description of available moves. For example, a pawn in chess moves one square forward along its line if unobstructed (or two on the first move), or one square diagonally forward when making a capture. This general description combined with a specific placement of pieces on the chessboard results in a number of available moves for each piece (that number being equal to or greater than zero) at a particular point in the game. The game theoretical decision tree that would be needed to chart the game of chess becomes unwieldy because of the number of possible configurations of the pieces on the board combined with the number of moves available to each piece. If we wanted to look at chess from a game theoretical perspective, we would need to simplify the game to a specific setting of pieces on the board and a very limited number of possible moves. For example, this specific pawn can either move one square or two squares forward from the square it is currently in. This would describe two moves available to this specific player. For the opposing player, we would have to make a similar description of available moves. We would then be able to make a decision tree combining moves and opposing moves, and identifying the advantages for the players of each possible combination. For analyzing the game of chess, this simplification would not be a useful exercise. To illustrate where its application lies, here is the classic example of a game theory problem, the prisoner's dilemma:

Two members of a criminal gang are arrested and imprisoned. Each prisoner is in solitary confinement with no means of speaking to or exchanging messages with the other. The police admit they don't have enough evidence to convict the pair on the principal charge. They plan to sentence both to a year in prison on a lesser charge. Simultaneously, the police offer each prisoner a Faustian bargain. If he testifies against his partner, he will go free while the partner will get three years in prison on the main charge. Oh, yes, there is a catch ... If both prisoners testify against each other, both will be sentenced to two years in jail.[11]

In this example, there are two moves available to each of the players. Either they keep silent or they testify against the other. The payoff of their move depends on the choice the other player makes. They know the moves he has at his disposal, but they do not know what his opposing move will be. The dilemma hinges on the payoffs of the four possible combinations of moves, which can be summarized in a two-by-two matrix (Table 5.2). In game theory, economic situations such as auctioning and bargaining are abstracted in the form of these types of games, after which an optimal strategy can be demonstrated.

The prisoner's dilemma can hardly be considered a game in the sense that I have given the concept in the preceding two chapters. It does not evoke play. Perhaps we could call prisoner's dilemma a very simple game of chance. But it is not a game that has a great deal of replay value, even overlooking the prison sentences involved. In fact, making a game theoretical representation of a 'real' game quickly becomes very complex. We can still do it with *tic-tac-toe*, but that is about as complex a game as will fit into a game theoretical decision tree. And *tic-tac-toe* is still a tedious game once both players have figured out the dominant strategy.[12] The interesting games, the ones that we talk about in this book – Monopoly, Chess, *World of Warcraft* – create a possibility space that is a lot bigger and is impossible to capture in a decision tree beforehand. The rules need to be set in motion in order to explore and observe the behavior that can emerge in this space.

Table 5.2   Moves and payoffs of the prisoner's dilemma

	Player 2 keeps silent	**Player 2 testifies**
Player 1 keeps silent	1 year, 1 year	3 years, 0 years
Player 1 testifies	0 years, 3 years	2 years, 2 years

Let me put the position that rules have in relation to games in a diagram. We can state that there are at least two layers to a game. At its core, a game is a set of rules. When put into motion, these rules result in gameplay. And in the case of video games, there is a layer in-between these two that communicates the rules to the player. Figure 5.1 shows these three layers.

The rule set forms the essential logical and mathematical structure of a game. The rules describe what is possible and impossible in this environment and how the game system reacts to actions performed by the player. This inner circle is the core of the design for a game, something we will cover in the next chapter. The rule set is communicated through the representation or declarative layer, which in the case of a video game is shown on the screen. This is another aspect of rules that distinguishes the video game from other games. Not only does the computer enforce the rules (as I mentioned before), it also communicates them to the player. Very rarely do video game players read the manual before starting to play a game, unless they are of my generation. Gamers will figure out the levitational qualities of the different types of birds they can put in their slingshot as they go along. Or they will learn about the different types of moves

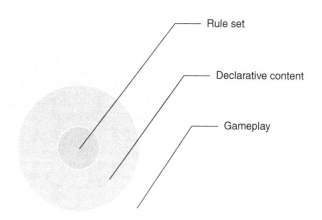

Figure 5.1   Three layers of video games[13]

and weapons in an introductory training level, without sacrificing any progress. The example of slinging birds shows another quality of rules in a video game. Because a video game creates a virtual world, the rules also need to describe certain inherent qualities of that world such as how gravity works. This is the case to a much lesser extent in board games and not at all in sports, where the rules build on the natural qualities of the environment in which the game takes place.

The outer circle in Figure 5.1 is the layer in which the actual behavior of the players takes place, which – as I mentioned before – cannot be fully anticipated in advance in a fully formed game. The behavior and experiences that take place in the outer circle are ultimately the result of the design of the inner circle, the rules. The patterns that emerge when these rules are set in motion are largely unpredictable and sometimes unintended or unwelcome.

At a recent London game design conference called Bit of Alright, these unpredictable patterns were on full display. At a level design workshop, the moderator gave participants one basic rule: you can only stand on paper, not on the bare floor. Based on this simple rule, the workshop participants needed to come up with game-like challenges.

One team modifies their game by tearing the paper to pieces and building very thin lines; later, they stick paper to vertical surfaces to create wall-running segments. Meanwhile, Marco's team has taken to writing 'L' and 'R' on each piece of paper. Without really expecting to, they introduce a new dynamic: put two right feet in a row and the player will need to hop.[14]

The insight that complex and unpredictable patterns can emerge from relatively simple rules is not unique to games. It is also an important element of complexity theory and its application to organizations.[15] This application focuses on the projection of the Complex Adaptive System (CAS) as identified in complexity theory on an organization and its environment. This proved to be a valuable addition to general systems theory, which traditionally addressed deterministic systems and attempted to capture them in a set of equations (as discussed in chapter 1). CAS are systems consisting of a large number of individual agents that interact based on rules and as such give the system as a whole an evolutionary and self-organizing quality.[16] Based on the CAS concept, complex systems can be formalized as a mathematical model. Besides this modeling approach, organizational complexity theory can also be approached empirically by matching a dynamic pattern to a series of observed organizational events.[17] Both approaches attempt to model the complex system in order to explain its behavior or explore possible future states.

There are two big differences between the inner workings of a CAS and those of a game. The first is that the rules in the case of a CAS are local. That is, each agent has his own rules. The second difference is that these local rules can evolve over time. These two qualities explain much of the evolutionary and self-organizing qualities that a CAS can have. In a game, rules have a universal and static quality. They are universal in the sense that every player has to abide by them. It may be possible that some roles in the game are allowed certain privileges or are limited in their options. The goalkeeper in soccer is the only player on the field who is allowed to touch the ball with his hands (inside the penalty area, that is). The priest character in *World of Warcraft* cannot acquire a permanent combat pet, whereas the hunter can. But the majority of the rules for a game are universal and all players have to submit to them. Even more important is the fact that the rules of a game are static. Changing the rules while a game is in progress will very likely get you into big trouble with the players. It may be necessary sometimes – like when players are gaming the system – but it is a delicate undertaking.

So we have seen that neither game theory nor complexity theory is suited for modeling games of the type we are discussing in this book. Yet when looking from the perspective of rules, we can see quite a few similarities between games and organizational systems. Both are characterized by a complex possibility space and both are prone to design flaws that can lead to undesirable behavior. So assuming that games are consciously designed – which they are, as will be discussed in the next chapter – we might ask if the approaches for designing games can offer a new direction for understanding and designing complex organizational systems. Before we follow that line of thinking further, there is one final, important distinction that needs to be made in the domain of rules.

5.3   Description, prescription, and circumscription

There is one important distinction with regards to rules that we have not made so far, but which is essential if we talk about them in the context of organizations. And that is the distinction between rules that are descriptive and rules that are prescriptive. In everyday use of the term, rules are considered prescriptive. They tell you what to do and what not to do. Descriptive rules are also possible. They are found in the natural sciences, where rules can be derived from observations in order to explain or predict phenomena. There, descriptive rules are usually called laws, with examples being the laws of thermodynamics or Newton's laws of motion. These types of rules pertain to inanimate objects, so there is not an issue of prescribing behavior (the stone would probably not do as you tell it anyway) but one of describing behavior under certain conditions. The rules are considered valid until someone has observed behavior that does not follow the rules. I apologize to natural science scholars who are reading this for not doing your field any justice, but this brief and incomplete description was simply meant to illustrate my wider argument, which is that one way to use rules is to describe, and possibly explain, a phenomenon.

Even though organization theory as a whole is generally regarded as a descriptive science – and sometimes attacked for its relevance as such – if we look at how organization theory has dealt with the concept of rules, it has mostly been in the prescriptive sense. There is extensive scholarship on the role that explicit and implicit rule making plays in institutionalization[18] and organizing processes,[19] which is a perspective I will not cover in this book. Another example of rules are those that agents in a CAS employ. They are prescriptive in the sense that they guide the actions of the agents. Extending this thinking to strategy, Stanford's Kathleen Eisenhardt has talked of a business strategy as being a set of simple rules that can guide a firm's behavior and prescribe how market opportunities should be pursued. She has shown a strategy of simple rules to be especially viable or even essential in unpredictable environments.[20] These simple rules are not just used by a firm's executives, but can be used to guide the actions of middle managers as well. Eisenhardt gives the example of (Danish hearing-aid manufacturer) Oticon's rule for canceling projects in development: 'if a key team member – manager or not – chooses to leave the project for another within the company, the project is killed'.[21] One can imagine quite a few ways to game the system which has this rule at its base, but that is not the point here. The point is that these types of rules prescribe a course of action.

Closely related to Eisenhardt's thinking on this subject is the concept of minimal structure.[22] It too sees a limited set of rules as a possible incarnation of the moderate amount of structure an organization needs in unpredictable environments. These rules of the minimal structure should guide the action of an organization while retaining its resilience.

Another way in which the concept of prescriptive rules has been introduced in the domain of organization and management theory is in the form of 'technological rules', proposed by Joan van Aken of Eindhoven University of Technology.[23] He draws a parallel with the relationship the descriptive natural sciences have with prescriptive applied fields such as engineering and medicine. The prescriptive rules or protocols to follow for building a bridge or curing a patient have a basis in the natural sciences as well as in rigorous research in their respective field. They are not the universal laws of the natural sciences, but they are rules tied to specific contextual variables. Given these symptoms and medical history, this would be the proper intervention. Van Aken proposes that organization and management theory focus more on producing these types of evidence-based rules, as guides for managerial problem solving. The types of rules that Van Aken is aiming for have a more general nature than those of Eisenhardt. They are tied to a certain field of application, but not to a specific organization. As an example, Van Aken derives the following rule from the work of Quinn[24]: 'If you want to realize a large-scale, complex strategic change, use a process of logical incrementalism.'[25]

A third and final incarnation of prescriptive rules in an organizational context is the concept of business rules, as used in information systems.[26] This is a type of prescriptive rule that comes very close to the rules Kathleen Eisenhardt describes. They are tied to a specific organization and prescribe a course of action if a particular situation arises. The difference is that these business rules are usually applied at the operational level, the level of business processes, and not at the strategic level as are Eisenhardt's rules. Business rules are normally used as a way to model a business process with the aim of automating it. In that context, an example of such a business rule can be: 'the record of a purchase order may not be entered if the customer's credit rating is not adequate.'[27]

These three views on prescriptive organizational rules are summarized in Table 5.3.

Table 5.3 Three types of prescriptive organizational rules[28]

Strategy as simple rules	Technological rules	Business rules
Used at the strategic and middle-management level	Used at the strategic and middle-management level	Used at the operational level
Tied to a particular organization	Tied to a specific context, but not a particular organization	Tied to a particular organization
Based on experience	Based on research	Based on experience

Now let us return to rules in the context of games. Are they prescriptive or descriptive? Let's take the game of *tic-tac-toe* as an example. If we assume that the 3 by 3 grid is already in place, then the rules are as follows:

1. Two players alternate turns.
2. When it is the player's turn, she places her symbol (either an X or a 0) in one of the empty cells.
3. The winner is the player who has placed three of her symbols in a row, either horizontally, vertically, or diagonally.
4. If all the cells are filled and there is no winner, the game ends in a draw.

I would argue that these rules are not prescriptive. They do not prescribe specific actions, but offer room for a variety of possible actions. The player does not know what the correct course of action for winning the game would be, based on these rules alone. But if you play the game, a strategy for winning – or rather, for not losing – will quickly emerge. I assume you know the one I am talking about. Since *tic-tac-toe* is a simple game – that is, its possibility space can be outlined – the strategy for not losing is failsafe. We could express that strategy in the form of rules (I will leave that exercise to the reader) and we would be back at the prescriptive rules of Eisenhardt. I would like to refer to these rules of thumb or heuristics for winning a game as a strategy.[29] Strategy in a business context is a subject I will return to in Chapter 7. But when I talk about rules in this book, I am not talking about heuristics or strategies. I am referring to the class of rules used in games. They are not prescriptive, but I do not feel comfortable calling them descriptive either. Let us call these rules circumscriptive. And where rules and circumscription may normally be associated with limitation, in games these terms should be associated with creation. The rules of a game create a new space inside which a vast variety of behaviors are possible. They circumscribe the framework inside which play takes place. In games, behavior is not prescribed by rules, but it is a reaction to those rules.

Applying this circumscriptive conception of rules to organizations opens up a new perspective. Identifying the mechanics of an organization and expressing them as rules could be a productive undertaking. These rules would circumscribe the possibility space for this particular organization. And organization should here be viewed broadly, as an arrangement that may cross the boundaries of individual firms. Taking this thinking one step further, we could then 'play out' the rules to increase our understanding of that possibility space. As I discussed earlier in this chapter, setting rules in motion is essential for understanding them. If we better understand the mechanics of the organization after having played with them, we may be in a better position to devise a course of action. Or alternatively, we could decide to change those organizational mechanics. These ideas will be the subject of the final two chapters of this book. To close this chapter, I will give an example of a project in which organizational rules played a pivotal role.

5.4   Playing with the rules or by the rules: We Beat The Mountain

WBTM is a company founded with the aim of designing, producing, and marketing products made entirely from recycled materials. At the time that this project took place the company was at the brink of bringing its first consumer product to market. Within this context, I was asked by the founder of WBTM to apply game-based methods to see if it could help them to further develop their strategy and way of working.

5.4.1   The story of We Beat The Mountain

The story of how WBTM was founded was ingrained in all the core team members, who related it almost identically. The founder got the idea when he noticed one of his suitcases had broken again, as it was coming down the luggage belt at an airport. The combination of his amazement at the bad design and poor materials of these suitcases (a reputable brand) and the need he felt to put his entrepreneurial skills to work in a socially responsible way led to the plan for WBTM. The first idea that came out of this experience was to produce a suitcase out of recycled car tires and from exploring that basic idea, the company took shape.

The core element around which the company evolved is its mission to beat the worldwide mountain of waste (which gave it its name). There are two main strategies that members of the core team mentioned for achieving this mission: one is to create and sell products made from recycled materials; the other is to look beyond the WBTM organization and to create a movement that would make people think differently about waste and recycling. For the short term, building the brand, selling products, and thus generating profit took priority because a steady cash flow was needed to allow the company to grow. This also meant that activities the organization undertook were mostly driven by opportunities that arose to meet these short-term goals. But the goal of creating a movement was ultimately considered more important in terms of contribution to the mission. The basic idea was to make it fashionable to use recycled products, so that people would start choosing what they buy differently. The goals of the organization had not been quantified in terms of numbers of products sold or amount of kilos recycled. As the founder said: 'I don't believe in rationalized business goals. I believe much more in these things taking a natural course.'

The organization that WBTM spun off from is called Kirkman Company, a consultancy firm for strategic sourcing, what they call 'Make, Buy or Ally' decisions. One of their core concepts is the so-called Flat World Company, an organization that only does the things it excels in and organizes its other activities through partnerships. Several core team members of WBTM – all very familiar with the Flat World Company concept – pointed out the irony of the fact that they tried to do everything themselves at the beginning, including the things they were not very good at. One of the reasons for this lay in the fact that there were insufficient funds to outsource activities. Several team members said that this lack of financial resources had impacted the speed at which WBTM could develop, specifically mentioning that they would be setting up several product lines at once if they had the funds. As WBTM evolved, more and more partnerships were established. It seemed the strategy for this was somewhat emergent. Outsourcing was considered when an area of activity started to cause problems. Several team members described WBTM as being the spider in the center of a web of connections with partners. When asked about the future, they described a small core team that directs all these relationships. But it was also apparent that none of the team members had given much thought to what the future organization would look like.

In the early stages of WBTM, everyone in the core team was doing everything. After some time, a clearer division of roles and respon-sibilities was achieved, although it did not reach the level of job descriptions. A clear distinction had been made between responsibili-ties for the front-end and those for the back-end of the organization. The front-end involved marketing and sales and the back-end was design and production. Despite this division, many team members still felt a responsibility for the entire scope of the organization. This was expressed in the fact that they were willing to take over activi-ties from others where necessary, but also in the fact that they felt the need to voice their opinion about a lot of subjects. There are two sides to this coin. On the one hand it was seen as a strong point that everyone could participate and have their say, but some team mem-bers also pointed out that these discussions could negatively impact the speed and quality of decision making.

Many team members mentioned the 'just-do-it' mentality that was instilled in WBTM by its founder, the freedom and confidence to start new initiatives and to see how far you get. The founder called it a strategic choice to try things out without exactly knowing beforehand what the outcome would be. But he combined it with the need to be open to feedback when things didn't go well. Other team members also mentioned these two sides: the freedom they got but also the fact that the founder would intervene when things were not moving in the right direction.

This way of thinking meant that the early stages of the organiza-tion were characterized by some as a 'journey of discovery'. All the core team members came from a consultancy background and had no experience in producing and selling products. One of the early lessons was that the starting point should be the raw material you are going to use for the product, not the design. This lesson was learned in the course of developing the original idea of a suitcase out of recycled car tires. Some early research showed that car tires are too heavy and very difficult to recycle because they consist of many different materials that are hard to separate. So they were not suitable as a raw material for a product. A second important lesson was that there is a big difference between designing a product and producing it. The first design for a laptop sleeve was the result of a design contest organized through the Internet. It soon came to light that this design was impossible to produce cost-effectively. The consequence of this lesson was that a partnership was formed with a local design firm that brought in the relevant expertise and could oversee the design and production process. As one team member said: 'We found out that even if you want to make a product in a "flat world" manner, you still need a little bit of knowledge about making products.' Despite the fact that the initial lack of knowledge meant that some things took longer than they could have, most team members valued these lessons and mentioned that there would probably be many more down the road.

5.4.2   The rules for beating the mountain

To help WBTM in developing its organization, I proposed an approach that consisted of four workshops with the end result being a validated organizational rule set. At the time we started this project, I envisioned the rule set to express WBTM's desired way of working and organizing, for the core organization as well as for its collaboration with business partners and the broader online community. In other words, it was intended to be prescriptive. It was to be a strategy, to use the terminology of the previous section.

After an initial round of interviews and the review of available documentation, a framework diagram for WBTM was developed. This framework contained a high-level description of the organization and indicated the main elements that had to be explored during the workshops in order to produce the rule set. It is shown in Figure 5.2.

The framework diagram expresses that there are two ways for WBTM to contribute to its mission. One is its core activity of producing products from recycled materials. But considering the magnitude of its mission – beating the worldwide mountain of waste – other contributions need to be made, which center on starting a 'movement', a different way of looking at waste and recycling. Both

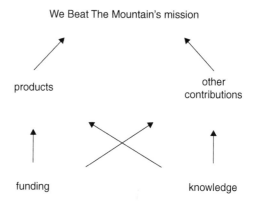

Figure 5.2   Framework diagram for WBTM

producing products and starting a movement required funding and the contribution of different areas of knowledge, not all present in the core team of WBTM. This framework contained a number of elements that had to be filled in, such as: Who are the stakeholders involved in WBTM? What are the areas of knowledge that these stakeholders can contribute? What are possible contributions that these stakeholders can make to WBTM as a movement? What can WBTM spend its money on?

To fill in these gaps, a number of workshops were conducted that used lusory spaces to generate the input needed. These workshops were comparable to the ones described as part of the ECC project in the previous chapter. Using all the information collected during these workshops, the framework was then developed into a set of rules that described the organization's mechanics. How this is done is something we will cover in the next chapter. The mechanics revolved around beating the mountain of waste by means of producing and selling products as well as by other contributions that the players could make. In the course of describing these mechanics, it became clear that some additional information had to be collected from the core team in order to make the rules most closely fit reality. To this end, an additional workshop with the core team was used. This workshop centered on classifying and valorizing some of the information collected earlier. For example, each contribution to the movement had to be connected to the goals of the players involved.

After the final missing information had been collected, the rule set was finished and visualized as a paper prototype. The why and how of 'paper prototyping' will be covered in the next chapter, but for now it is sufficient to know that the prototype took the form of a board game, using poker chips, scoring forms and cards that participants could use at certain moments in the game. The goal of the game was to beat the mountain of waste within a set time limit. The winner would be the player who was able to obtain the highest score on their individual goals. But if the overall goal of beating the mountain was not achieved, there would be no individual winner either.

After testing it in a workshop, the rules of the prototype combined with the observations during the workshop were used to produce a validated organizational rule set. This was presented as the end result of the process in a session with the founder of WBTM. One example of these rules was:

> The players do not form a network with We Beat The Mountain as the central node, but there are bilateral collaborations between the players that should be encouraged by We Beat The Mountain.

5.4.3 Rules on two levels

On one level, rules were used in this project as a way to structure the individual workshops and to create lusory spaces inside of them (see Figure 5.3). This meant the typical elements of gameplay could be witnessed: an eagerness to participate, competition, zeal for winning and an overall atmosphere of energy and fun. Although the majority of the participants of the workshops accepted and adhered to the rules, there were also attempts by some of the workshop participants to circumvent or manipulate them. This is not surprising, since mastering, beating, and even subverting rules is a part of gameplay.[30] In some workshop participants, the imposition of rules seemed to invoke an urge to explore the boundaries of the system that the rules circumscribe, to play *with* the rules, rather than *by* the rules. They were keen on finding loopholes and shortcuts in the rules that allowed them to turn certain aspects of the workshops upside down. In other words, they were trying to game the system. This could indicate that some of the participants were more interested in playing with the game system as whole – and as such, with the game masters – than in playing the game that the game masters had in

Figure 5.3   A scene from one of the workshops at WBTM

mind. This testing of the boundaries increased in relevance during the last workshop, when the rules circumscribed the organizational system in question. Testing the limits of the rules then became a way to show certain, possibly undesirable, organizational dynamics more explicitly. It elucidated the consequences of playing by certain rules and thus the mechanics of the organization.

When asked about this behavior afterwards, workshop participants gave it different labels: cheating, manipulation, or sabotage. In discussing the reasons for this manipulation (the most common term used), three reasons come to the fore. The first is achieving an individual advantage, which is an obvious reason for cheating (as discussed earlier in this chapter). The other two reasons can be seen as variations of gaming the system. One of those lay in testing the limits of the rules, as part of a mindset or personal impulse to challenge existing conventions. It is important to realize that not everyone will accept rules without protest, even if they exist within the context of a game. The passive aggressive spoilsport is never far away. Questions about the objectives and the outcome of what they are doing will come up in participants' minds and will sometimes be voiced, depending on their personality or position in the group. The second variation on gaming the system was challenging the rules to make it clear to other players what the consequences of these rules are. As the chief manipulator of the rules commented:

> I'll play the game the way we agreed, that's fine by me. But with a realization that if we're aiming for individual gain, then at least deliberately so.

Manipulating the game is not the most common behavior, however. Most participants are eager to play the game within the confines and the spirit of the rules, and their frequent questions to the game masters are meant to clarify the rules, not to challenge them.

So on one level, the rules acted as a way to structure the workshops. But as I mentioned earlier, an organizational rule set was also the intended end result of this project. This rule set was to be prescriptive of the future way of working and organizing of WBTM. When asked about the outcomes of this project afterwards, members of the core team mentioned three elements. The first was the journey itself; the reflection and awareness the workshops had created among the core team of WBTM.

> Maybe in that sense the process is even more valuable than the exact outcome. What I mean is the awareness, deliberately contemplating: how will I deal with my surroundings, with the players.

The concrete, tangible result was the second element mentioned. This consisted of the game (the 'paper prototype') that was developed, which was described by a member of the core team as follows:

> The result is that we have now developed a We Beat The Mountain globe that we can travel across. So we know how the interactions work and that there are many possible destinations.

A second part of the result was the set of organizational rules for further developing WBTM.

> That [the rule set] contained some recommendations that you recorded during the journey which are useful for us in our reflection about our day-to-day operation.

If we look at these comments, we can see that the value of the rule set for this organization does not lie in prescription or strategy. As is the case with a game, the rules developed here circumscribe a possibility space. They show how WBTM operates and how it interacts with the stakeholders in its environment. What's more, it was not so much the rules themselves that were valuable, written down in a report at the end of the project. It was playing with those rules in the form of a 'paper prototype' that created awareness of the organization's mechanics and proved to be an important learning experience for those involved. This 'paper prototyping' and 'playtesting' are part of the toolkit of the game designer, which will be the subject of the next chapter.

6
Design

From 1998 to 2002 Fred Collopy and Richard Boland, professors at the Weatherhead School of Management at Case Western Reserve University, had the privilege to work with architect Frank Gehry on the design of their new building. Gehry is generally considered one of the most important architects of our time, whose iconic and innovative buildings such as the Guggenheim Museum in Bilbao draw visitors from all over the world. Boland and Collopy's collaboration with the architect allowed them to experience first-hand how he and his partners work and approach problems. It was an inspirational episode for them, which opened their eyes to a very different mindset from the one they were instilling in their management students. From working with Frank Gehry, Boland and Collopy developed the view that 'if managers adopted a design attitude, the world of business would be different and better'.[1] At the occasion of the opening of the Peter B. Lewis Building in June 2002, they assembled scholars, artists (Frank Gehry among them) and managers for a workshop entitled 'Managing as Designing'. This workshop and the book that resulted from it[2] were very influential in casting a new light on organizational design,[3] an academic subject that until then seemed to have gotten stuck somewhere halfway through the twentieth century.

6.1   The evolution of organizational design

Speaking about an organization as something that can be designed can seem problematic. The prominence of human actors in all organizational endeavors makes it hard to draw parallels with traditional design fields such as architecture or engineering. However, there have been attempts in organization theory to support managers and consultants in their role as organizational designer, as the person who chooses or shapes (more about that distinction later) the structures and processes that allow an organization to operate effectively. Organizational design views the manager as a designer and management theory as a design science, which Nobel laureate Herbert Simon has described as 'a body of intellectually tough, analytic, partly formalizable, partly empirical, teachable doctrine about the design process'.[4]

Organizational design can be understood in terms of three waves, as presented below: the rational systems perspective, contingency theory, and the design attitude.

6.1.1　The first wave: The rational systems perspective

The pioneers in organization studies sought to develop general principles that would be applicable to organizations. Particularly administrative theorists at the beginning of the twentieth century such as Henri Fayol developed guidelines for managerial decision-making. Theirs was a view – inspired by engineering – of the organization as a unified and formalized machine that can be deliberately designed as such. In this first wave of organizational design, the artifact to be designed was the formal structure of the organization, consisting of such elements as the organization chart, role requirements, and procedures. The underlying idea was that the brainpower was concentrated with the managers at the top of the organization, and that their designs and instructions would be adopted and followed by the rank and file.

6.1.2　The second wave: Contingency theory

In the mid-twentieth century, there was a realization among organizational scholars that there are no general rules for how to design an organization. As they observed different organizational designs in the field, they began to ask themselves why there was such a large variety of apparently successful designs. The answer lay in the fit between the design chosen and the environment that the organization was operating in. This led to an approach that emphasizes that design decisions depend – are contingent – on environmental conditions or the type of organization they are applied to. This is the essence of Paul Lawrence and Jay Lorsch's contingency theory,[5] which can be understood as the second wave in organizational design. The contingency school showed a promise of useful implications for managers and has been a dominant approach to organizational design throughout the second part of the twentieth century, with influential work such as that of Jay Galbraith,[6] David Nadler and Michael Tushman,[7] and Michael Goold and Andrew Campbell.[8] Henry Mintzberg's five basic design configurations (simple structure, machine bureaucracy, professional bureaucracy, divisionalized form, and adhocracy) have also been an influential model in the contingency theory school, with the rationale being that certain contextual factors determine the choice for one of these organizational designs.[9] At the same time, it was becoming clear that a contingency theory approach was not sufficient, because the emphasis remained on structure and control, business processes and job descriptions. But close inspection by organizational scholars revealed structure to be an elusive concept and one that may not be a productive object of design.[10] There exists a tenuous relationship between what an organizational designer comes up with at his drawing board and what actually transpires in the organization where the design is implemented. A radical re-assessment of the field of organizational design was necessary to prevent it from losing its relevance.

6.1.3 Organizational design's third wave: The design attitude

Boland and Collopy's 'Managing as Designing' introduced that fundamental new perspective, which has been pursued over the past ten years. Inspired by fields such as industrial design and architecture, a unique mindset and approach to problem solving is presented as a guide for managers. Boland and Collopy call it the 'design atttitude'. Others, such as Heather Fraser at the Rotman School of Management[11] and Tim Brown,[12] CEO of design firm IDEO, talked about 'design thinking'. These ideas about how a designer thinks and works were not new. It was their application to business endeavors that was.

When trying to understand how a designer approaches problems, the work of design researcher Nigel Cross is very helpful. He has spent the last thirty years teasing out what exactly are the 'designerly ways of knowing', based both on laboratory work and theoretical reflections. Cross has identified four elements of design ability[13]:

- *Resolving ill-defined problems.* Because design problems are inherently ill-defined and complicated, the scientist's or manager's attempt to fully understand it is ineffective for reaching an appropriate solution in time.
- *Adopting solution-focusing strategies.* Instead of spending a lot of time with the problem by analyzing it or trying to fit it into a model, designers spend most of their time on generating and testing potential solutions.
- *Employing abductive or productive reasoning.* By working on potential solutions, designers focus on what may be, not on what must be (deduction) or what actually is (induction). This abductive or conjectural logic is essential in the work and mindset of a designer.
- *Using non-verbal modeling media.* Drawings, scale-models, and prototypes characterize the way designers communicate about their process.

As Boland and Collopy started to explore in earnest the application of these design abilities to management, they contrasted the 'design attitude' with the prevailing 'decision attitude' in management practice and education.

> The decision attitude assumes it is easy to come up with alternatives to consider, but difficult to choose among them. The design attitude toward problem solving, in contrast, assumes that it is difficult to design a good alternative, but once you have developed a truly great one, the decision about which alternative to select becomes trivial.[14]

From this perspective, organizational design becomes a process of shaping rather than choosing. This becomes all the more important since there may not exist an organizational structure or process on the 'solution shelf' able to cope with the current contingencies that managers are faced with, as discussed in Chapter 2. Several management scholars have been calling attention to the fact that most accepted academic theories of organization and management are based on research conducted at least twenty years ago,[15] when the circumstances that organizations were dealing with were vastly different. Scholars of organizational design such as Georges Romme have called for new ideas and new ways to solve problems.[16] A solution-focusing, conjectural organizational design approach may be such a way forward in these radically changing circumstances.

Organizational theorist Karl Weick was one of the keynote speakers at the 2002 'Managing as Designing' workshop that Richard Boland and Fred Collopy convened. There, he made two important observations in contrasting the design attitude with normal management practice.[17] One is that managers tend to 'overdesign'. Instead of working from the premise that behavior cannot be fully predicted or designed, they tend to get down to the level of structures and procedures. The motivational costs of this overspecification by managers are well documented.[18] He contrasted this to the process that Frank Gehry follows, who spoke about this earlier during the same workshop.[19] Gehry sees his architecture as playing a role in creating desired interactions. He is not trying to design these interactions, but he is simply aware of the effect that the design of his building can have on the behavior of its users.

I am reminded of an example from my own professional practice. In 2009, I was involved in a project to introduce a new way of working to the headquarters of a multinational insurance company. In this project I collaborated with the Dutch architect Ferdinand van Dam on the interior design of the building. The goals of the project were to increase collaboration between the different functional departments, to encourage knowledge sharing between employees, and to use the office space in a more flexible way. The abductive, iterative way of working that Ferdinand had was refreshing, albeit occasionally difficult to combine with the need for specifications and schedules of the client. One of the design problems that we had was creating a cafeteria that could double as an informal meeting space. Based on a short design brief, Ferdinand went ahead and created a stunning space that contained a variety of seating arrangements. There were traditional tables with long wooden benches for groups to sit at, as well as round white tables. There were semi-secluded diner seats inside the cafeteria as well as café-style tables 'outside', on the edge of the bright atrium. There was low seating as well as higher tables with stools round them. After construction had been completed, it didn't take long for people to flock to the new cafeteria

(Figure 6.1), not only during lunch hours, but throughout the day. It became the building's favorite meeting space, always buzzing with activity and with people running into each other, enquiring about what they were working on or arranging to join forces on a project. The enormous success of this space was unexpected (except perhaps by Ferdinand) and it became the best embodiment of what we were aiming for in this project. No manager advised these people to have their meetings here. No one even designated this as a meeting space. It was purely the attraction of a functional, pleasant, and aesthetically pleasing configuration of furniture, materials, light, and other elements that created these interactions.

Karl Weick stated that if managers were to 'underspecify' – as architects such as Frank Gehry and Ferdinand van Dam do with regards to the interactions in their buildings – it would increase the vitality of the organizational design. It would give the employees on 'the frontline' the freedom to react to what was actually happening in the field, instead of blindly following procedures thought up by people who may never have interacted with an actual client. If managers want to make full use of the skills of their employees, it is counter-productive to organize their tasks in such a detailed way that they are forced to turn off their brains.

Figure 6.1   The cafeteria in question (OTH Architects, 2010)

Apart from the somewhat elusive notion of 'design attitude', this new way of looking at organizational design also has implications for the appropriate organizational design *process* as well as the design artifacts that are the result of that process. With regard to the first aspect, the recognition of a design attitude has meant linking organizational design processes more closely to action and implementation. The use of models and prototypes that go through several iterations is a common feature of what a designer does. Boland and Collopy have called for a deeper integration of these types of practices into the organizational design process. But this action-orientation may also mean that a design is not finished after it has been realized and implemented. There is a need for constant improvement based on feedback by its users and for adaptation to changing circumstances.

In the case of an industrial designer or an architect, the artifact that is the result of the design process is a representation or specification of the object to be produced. In the case of the organizational designer, that relationship is more problematic since what actually takes place in an organization may have just a loose connection with what was specified as the organizational design.[20] Traditionally, the artifact that results from an organizational design process would take the form of an organization diagram, job description, or a business process flow chart. Based on the discussion in this chapter so far, we can deduce that the designed artifact should be of limited specification or at the very least retain vitality. If it is an underspecified artifact that we are after, then we need to be concerned with how to accomplish this. Let me return to the rules we talked about in Chapter 5 – not the prescriptive strategy rules, but the circumscriptive rules that form the core of a game. This rule set seems to be an artifact of limited specification, which nonetheless lets complex behavior emerge. It is therefore useful to take a closer look at the design process of such a rule set.

6.2 Video game design

Let me first clear up a common misconception about video game design. The role of video game designer is not the same as that of video game developer. The latter is the more technical role, the person who does the programming, the graphics, or the audio. Depending on the size of the production, these can be separate, highly specialized teams. The game designer is the person who envisions the gameplay or the experience of the player that he or she would like to see and then designs the high-level framework within which that gameplay will take place. That framework is the rule set. Of course, in small projects the roles of game designer and game developer can be combined in one person. But conceptually, it is important to distinguish between the two. Game designers may even do their design work independently from the medium in which the game will eventually be presented to the player (board game or video game), although that is not very common.

In Figure 5.1 the three layers of a game were depicted; the visible behavior called gameplay as the outer circle, the rule set as the game's core, and a content layer in-between. The game designer is concerned with the outer circle, the behavior and experiences of the player. What happens in this outer circle is ultimately the result of the design of the inner circle, the rules. As we have seen in the previous chapter, rules often have unexpected consequences. It is not possible to fully anticipate what will happen once the rules are set in motion. Game designers will develop an intuition about the gameplay that certain rules will give rise to, but a test will always be necessary.

Rob Pardo, the lead designer of *World of Warcraft* and currently the Chief Creative Officer at Blizzard Entertainment, gave a very good example of this need to test during a talk at the Game Developers Conference in 2008, the largest annual gathering of the industry. Pardo gave the example of Alterac Valley. Alterac Valley is a so-called player-versus-player battleground that is part of *World of Warcraft*. *World of Warcraft* can be played in two modes: player versus environment (PvE) or player versus player (PvP). The former is the most common, in which the only damage you can sustain is caused by the game system. Other players cannot hurt you. They can humiliate you by beating you in a duel, but that will only hurt your pride, not your game character's health. In PvP, you are fighting other players. Battlegrounds are part of PvP play, in which players for each of the two factions of the game (the Horde and the Alliance) form teams of forty and clash in large-scale battles. When Blizzard designed Alterac Valley, it intended it to be a massive PvP melee where each team would rush into the middle of the Valley and meet in an epic battle between Horde and Alliance, culminating in the elimination of the opposing team's general. What actually happened once players figured out how the battle could be won was that both sides would run right past each other in order to storm the opposing faction's base to see who could kill the general first. Pardo said that if there would have been a way for the game characters to wave at each other while storming by on their mounts, they would have done so while shouting 'Have fun storming our castle!'.[21] It was an example of the behavior I called 'gaming the system' in the previous chapter. Players of a game such as *World of Warcraft* will be ruthless in bringing your design flaws to light.

6.2.1 The video game design process

Game designers are tackling what Katie Salen and Eric Zimmerman call a second-order design problem.[22] They realize that they cannot directly design player behavior, so they do it indirectly through the design of the rules of the game. This also implies an iterative approach, in which rules are designed, set in motion, the resulting gameplay is observed, the rules are adjusted to prevent undesirable behavior, and the rules are set into motion again (Figure 6.2).

Game designer and educator Tracy Fullerton gave a clear description of this iterative process in her book on game design.[23] Her description now forms the basis of many video game design curricula. It captures the second-order design problem in that it has the player experience as a goal and the rule set as the object of design. Through a prototyping process involving the players, there is a constant evaluation of the design goals. The process consists of five steps, represented in Figure 6.3 and explained below.

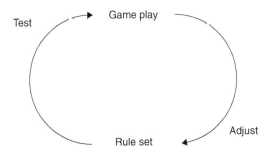

Figure 6.2   Second-order video game design

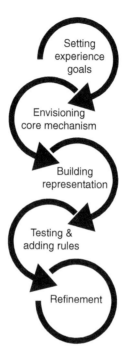

Figure 6.3   Steps in the video game design process[24]

1. *Setting experience goals.* Game designer Katie Salen describes this as follows: 'Most often I begin by trying to define exactly what it is I want a player to experience – how I want them to feel, what physical movements or actions I want them to enact, in what ways they might interact with other players or contexts.'[25]
2. *Envisioning the core mechanisms.* In this step the core mechanisms that will have the experience goal(s) as an outcome need to be envisioned. A core mechanism is an action that the players repeat often, so it should be described in terms of verbs. Tracy Fullerton uses a core mechanism from the game of Monopoly as an example: 'Players buy and improve properties with the goals of charging rent to other players who land on them in the course of play.'[26]
3. *Building a representation of the core mechanisms.* This step involves prototyping the basic objects and key procedures that are involved in the core mechanisms. Examples of objects in games are the

king in chess or the bank in Monopoly. The key procedures are the most important actions that the objects can perform in a certain state. Their available moves, so to speak. The objects and their properties, as well as the procedures should then be represented. This is called paper prototyping and can be done with simple materials such as cardboard and paper. The core mechanisms will then more or less take the form of a board game.

4. *Testing and adding rules.* After the core mechanisms have been represented as a paper prototype, this prototype should be rigorously tested. This is done in so-called 'playtesting' sessions, in which users play the game under the supervision of the designers. At some points during the initial round of playtesting, the game will not be able to proceed because of questions about which actions are allowed and which effects actions have. This can be overcome by adding rules that answer these questions based on certain conditions. In this part of the process the rule set takes shape.

5. *Refinement.* At this stage, there is a paper prototype that can be played independently of the designers. The goal of the refinement stage is to make sure that what was designed meets the original experience goals. If this is not yet the case, the game should be balanced. This can be done by manipulating the object properties or changing rules that cause undesirable dynamics.

This last step of refinement and balancing is very important because many games show undesirable dynamics in their first iteration. What often happens is that there is a dominant short-cut strategy for winning the game or that players pull too far apart early on and it becomes impossible for those in pursuit to still win. In fact, one could say that the game of Monopoly has a balancing issue because what often happens is that the rich keep getting richer (which may have been the main player experience the designer was aiming for). If you own some strategically placed property in the game, the money will start coming in, which allows you to buy more houses and receive even more rent. This is what Tracy Fullerton calls a reinforcing relationship,[27] which may create an imbalance in a game.

6.2.2 Meaningful play

If we talk about video game design, we should not limit ourselves to a description of the process. It is also important to talk about some of the values that have become embedded in the video game design community concerning what constitutes a good game. Most recently, these values have come to the fore in the discussions surrounding 'gamification', a phenomenon that has been dismissed by many game designers as not getting at the core of what games are (see chapter 3). One of the ways to label the core of games is by using the concept of 'meaningful play'. In well-designed games we see autonomous individuals devising short and longer-term strategies, reacting to changing situations, and processing the information needed to complete their task at a fast rate. This behavior is what happens when meaningful play occurs, which has been described by Salen and Zimmerman as the goal of successful game design.[28] They identify two criteria for meaningful play, which have also been mentioned by Doug Church[29] and Tracy Fullerton[30] as important elements in video game design: 'discernability' and integration.

Discernability. This means giving direct, clear feedback to players about the result of their action. Church calls this perceivable con-sequence. Salen and Zimmerman give the example of shooting an asteroid in a video game. The system needs to give the player some kind of visual, aural, or tactile feedback if you hit the asteroid. In multiplayer games, this feedback can come both from the game system and from the other players.

Integration. This principle revolves around letting players know how their actions will affect the rest of the game so they can make a plan and then act on it. Church calls this intention. Fullerton mentions two different principles that together make up intention: one is presenting players with meaningful choices; the other is giving them clear and focused goals.

Meaningful choice is a concept that is central to understanding what makes a game interesting. In a game of *tic-tac-toe*, there are no meaningful choices. If you want to avoid losing, there is usually only one possible choice of where to place your X or 0. To make a game more interesting, there needs to be a more complex trade-off between possible moves. For instance, if a game supplies me with a powerful weapon that I can only fire once, I can choose to use it right now to easily blast away this enemy I have in front of me, or I can save it for the boss battle that I know is coming at the end of

Figure 6.4   Meaningful play

this level. There are many patterns that can be used to create more meaningful choices in games. Staffan Björk and Jussi Holopainen have done valuable work in collecting some of these game design patterns.[31] The example I just gave is what they would call an 'action cap' (you can only fire the weapon once). The core of the criticism leveled at gamification initiatives often centered on their lack of meaningful choices. If I can win a game just by checking in many times to a certain location or by clicking many times on a website button, the gameplay likely offers very little fulfillment. But if I survive a level by a combination of skill and choosing the right strategy, I may experience this gameplay activity as intrinsically motivating.

There is a third principle of meaningful play, besides discernability and integration, which is not found in the literature on video game design, but which needs to be made explicit to understand players' behavior. That is the principle of recoverable loss.

Recoverable loss. In *World of Warcraft* and in most other contemporary video games, what you risk by failing is relatively easy to recover. In the case of *World of Warcraft*, your ghost will be resurrected at the graveyard after you die and you need to do a so-called 'corpse run' back to your game character's dead body. You can then continue the game without having lost any points or gear. You have just wasted some time. Combined with discernability and integration, recoverable loss becomes a powerful principle, which game designer Doug Church describes as follows:

Players get to make a plan, try it out, and see the results as the game reacts. And since that reaction made sense, they can, if needed, make another plan using the information learned during the first attempt.[32]

This trial-and-error cycle of meaningful play is depicted in Figure 6.4.

6.3 Applying the video game design process to an organization

We are now nearing the core of my argument. Earlier in this chapter I explained the push in organizational design towards a design attitude for managers and suggested that the artifact to be designed in that new approach could be a set of circumscriptive rules, such as the ones found at the core of a game. In the previous section I explained what the video game design process, which has that rule set as its objective, looks like. What I'd like to do now is make the case for applying this video game design process to an organization as a form of organizational design.

In order to make the application of the video game design process to organizations possible, I decided that some changes to the approach were necessary. Some of these changes were made based on theoretical insights. Others were a result of the first application of the methodology, which was at the ECC described in chapter 4. One of these changes is the start of the design process. The starting point is not setting experience goals, as in the case of video game design, but establishing the goal, mission, or – to use a term from the field of video games – the 'epic meaning' of the organizational endeavor that is under design.[33] In video game design, the second step would be to envision the core mechanisms that have the experience goals as an outcome. In the organizational context, this step is further elaborated. First, a high-level framework is established that describes how the organizational system contributes to its 'epic meaning', based on interviews with key informants. Figure 5.2 is an example of such a framework. Then, through a number of workshops, this framework is filled out and the core mechanisms are envisioned. Core mechanisms in this case are courses of action directed at achieving the goals or epic meaning.

The third step in video game design is building a paper prototype that represents the core mechanisms. This is a crucial step in the design process because it is the part where the magic happens, so to speak. Here the game has to start taking shape. In the organizational application of the process, this is no different. In this step, with the exploration of the framework as the input, a game structure needs to materialize and needs to be expressed in the form of rules. Working from a framework turns out to be essential here. At the ECC, we did not do this and started the design process with a very open brainstorm about players, locations, activities, and other components. For the paper prototype we used an existing game mechanics framework. We then found out that the results of the brainstorms led to an uneven playing field with unbalanced components. It proved necessary to have a framework for the prototype first and to let that framework direct the brainstorms. Even when using this framework, the game will not materialize automatically. This step of building the paper prototype requires looking at an organization through a game design lens. There is very little I can add to make this part of the process more explicit. What helps is closely studying how a game designer works,[34] looking at game design patterns such as the ones documented by Björk and Holopainen,[35] and of course playing games.

After the paper prototype has been built, what follows in video game design are several rounds of playtesting. The first rounds have the focus of adding rules and are done by the designers or under their close supervision. The later rounds of playtesting focus on balancing the game. By then, the paper prototype should have reached a stage in which it can be played independently by the players. In the organizational application of this process, playtesting follows a similar pattern. The first iteration of the prototype is tested by the designers to find out if it is actually playable and to bring it to a level that sufficiently reflects the organizational system under design. This goal of the prototype reflecting the organizational reality will very likely lead to a need for additional information from the key informants.

This is a crucial point of divergence between the organizational design process and the video game design process. The latter deals with the creation of a completely new set of rules, whereas the rules of the former need to reflect an existing organizational reality. It may be that we are aiming to make changes to the organization through this process (which is something I will get back to later) but initially the paper prototype needs to be an expression of the current organizational rules. This is another lesson we learned during the ECC project. There, we had originally planned to do a design workshop with the core client team after finishing the first version of the paper prototype, but we ended up canceling it. This meant that we were not able to answer all the questions that arose during our design and initial playtesting of the prototype. One of the ways we solved this was by letting the players fill in some of the blanks at the start of their playtesting session. However, this was not completely to our satisfaction. It proved necessary to involve the organizational experts (in this case: the core client team) more closely in the paper prototyping stage of the design process. After this additional information has been incorporated, the paper prototype should be finished to a level that makes it playable independent from the designers. The final step is then a playtesting session in which the paper prototype is played by representatives of the different stakeholders that have a role in this organizational system. Table 6.1 compares the video game design process to its application in an organizational context.

Table 6.1   The video game design process applied to an organization

Video game design process		Organizational application	
1.	Setting experience goals	1.	Setting goals and establishing framework
2.	Envisioning core mechanisms	2a.	Filling out the framework
		2b.	Envisioning core mechanisms
3.	Building paper prototype	3.	Building paper prototype
4.	Playtesting, round 1: adding rules	4a.	Playtesting, round 1: adding rules
		4b.	Obtaining additional information
5.	Playtesting, round 2	5.	Playtesting, round 2

6.4 Who is the organizational designer?

In the organizational application of the video game design process as described in the previous section, there is a need for someone who directs the process. Let's call this the organizational designer. The organizational designer can be one person, but it can also be a team of two or even more people. Regardless of whether it is a solo or a team effort, I will speak of the organizational designer in this text. There is an important difference between an organizational designer and a game designer, apart from the obvious distinction that the game designer is not designing an organization. The difference is that the organizational designer in this case is not the person who shapes the organization. She is not the one who comes up with the creative ideas for new designs the way a game designer or an architect might. So in fact, there is very little abductive reasoning by the organizational designer. Simply put, the organizational designer in this case is directing a highly structured design process in which she puts the pieces of the puzzle that are handed to her by co-designers (organizational actors) in their right place. By looking a little closer at the ECC and WBTM projects, we can identify a number of elements that the organizational designer brings to the table which are important for the course and the outcome of the design process:

- An external view that allows for the identification of a framework that forms the basis of the design process.
- The role of moderator or 'game master' during a number of workshops, which entails designing and implementing lusory spaces that encourage creativity and an open exchange of ideas to collect the necessary information.
- The ability to design and balance a rule set that reflects the organizational system and to build a paper prototype of that rule set.

The design process is a process of co-creation together with the members of the organization in question. There are actually two circles of co-creators. There is an inner circle formed by a core team (three or four people) of co-designers, who are the primary source of domain knowledge about this organizational system. And then there is an outer circle of stakeholders, who supply information from their specific perspectives. Table 6.2 shows in which stage of the process

Table 6.2   Actors involved in the design process

Stage of the design process	Actors involved
1. Setting goals and establishing framework	Organizational designer, co-designers
2a. Filling out the framework	Organizational designer, co-designers
2b. Envisioning core mechanisms	Organizational designer, co-designers, stakeholders
3. Building paper prototype	Organizational designer
4a. Playtesting round 1: adding rules	Organizational designer
4b. Obtaining additional information	Organizational designer, co-designers
5. Playtesting round 2	Organizational designer, co-designers, stakeholders

the organizational designer, the co-designers and the stakeholders are involved.

6.5   What happens during game based organization design?

In this chapter, I looked at the current state of organizational design and observed that there is a need for practices that incorporate forms of design ability such as an iterative approach and the use of prototypes. I also noted a need for organizational design artifacts of limited specification and speculated that a rule set might be such an artifact. I then went on to argue that applying the video game design process to an organization may be a way forward in this context and I explained what this process would look like, based on actual applications of this idea in the field. I will label this process as 'game based organization design' from now on. So what exactly do we see happening during game based organization design?

6.5.1   Increasing the understanding of an organizational system

The premise in this chapter has been that a rule set was the artifact that we were after in this design process. But if we look closely at the process that takes place if we apply video game design to an organization, it is the uncovering of an existing organizational rule set rather than the designing of a new one. The implicit assumption at the start of game based organization design is that the stakeholders individually do not fully understand the organizational system they are part of. I would argue this is a safe assumption to make in the current, turbulent environment. To increase this understanding, game based organization design focuses on two parallel interventions. The first is to combine the knowledge and perspectives that the different stakeholders have of the organizational system, to combine the pieces of the puzzle, so to speak. And the second intervention is to let stakeholders experience the organizational system in action. The first intervention (combining the knowledge and perspectives of the stakeholders) is achieved by creating a number of lusory spaces (see Chapter 4) in which creativity and an open discussion climate are encouraged. The lusory spaces are used to brainstorm about different elements of the organizational system (directed by the framework that was set up at the start of the process) and about possible courses of action to achieve the goals (also identified at the start). In and of itself, these lusory spaces form a powerful organizational intervention that yield bonding and shared understanding between the stakeholders, and produce a great number of creative ideas. But with regards to these outcomes, the intervention does not differ much from other interactive organizational interventions such as appreciative inquiry,[36] future search,[37] and open space technology.[38] However, the information and ideas that are the result of this first intervention are the input for the second intervention, experiencing the organizational system. In this second intervention, the rules and the game come into play. What happens is that the organizational designer builds a prototype of a game based on the information about the organizational system that he or she has been able to uncover with the help of the co-designers and the stakeholders. As discussed in this chapter, building a game entails designing a set of rules. But unlike in game design, the rule set in game based organization design is not the end result of the process. The rule set becomes a temporary, intermediate artifact that is expressed and evaluated in the form of a paper prototype. These rules could later be extracted from the prototype game and written down, which is something I have done in projects such as the one at WBTM (see Chapter 5). But as soon as they are put in writing, they are in danger of losing much of their value. The value of this part of the intervention lies in the experience of playing with the rules in the form of a paper prototype and finding out their possible effects. The result is a measure of awareness for stakeholders about the constituting elements of their organizational system, their interactions, and the possible outcomes of this dynamic. In the terms of the previous chapter, the stakeholders are exploring a possibility space circumscribed by rules and increasing their understanding of it.

6.5.2 Iteration and abductive reasoning

Earlier in this chapter, I discussed a number of attributes of the design attitude. One is the use of non-verbal modeling media, which game based organization design employs in the form of the paper prototype. There are two other attributes of the design attitude which merit attention in relation to game based organization design. One is the iterative way of working and the other is abductive reasoning (focusing on what may be, instead of what is).

Iteration in game based organization design is not so much building a prototype, testing it, and improving upon it – although that is also an aspect of the process – but more importantly creating progressive understanding for organizational actors about the system they inhabit. According to consultant and educator Jamshid Gharajedaghi, the holistic enquiry that is such a vital part of systems thinking requires understanding of structure, function, and process at the same time.[39] To this end, he describes an iterative, progressive process of inquiry. Game based organization design also follows this iterative mold. It starts by taking stock of an overarching goal ('function' in Gharajedaghi's terms) and a framework that is in place to reach that goal ('structure' and 'process' in Gharajedaghi's terms). It then proceeds to add detail to this framework in focused brainstorm sessions and by having the different stakeholders explore how they can contribute to the overarching goal. This then culminates in the design of a rule set and the playtesting of this rule set in the form of a paper prototype. This 'journey' (as participants have described it) of experiential learning progressively increases the holistic understanding that the participants have of the organizational system they are part of. The iterative nature lies in using each step of game based organization design to build upon the knowledge collected during the previous step. The co-designers and stakeholders encounter and revisit the building blocks they themselves contributed in an earlier stage of the process.

Promoting abductive reasoning was one of my original goals in applying the video game design process to organizations. But the way game based organization design plays out, its focus lies on the organizational system that is, not on the system that can be. Abductive reasoning does take place during the process. This happens when stakeholders generate new courses of action (called core mechanisms in the approach) to contribute to an overarching organizational goal. But where it pertains to the organization itself, the reasoning is more inductive than abductive. Inductive reasoning is very helpful for participants to gain a deeper understanding of their organizational system, but it does not generate entirely new designs. In the course of the process, an organizational system is circumscribed and explored. But we stay inside that circumscription. That is an important difference between this process and the design processes it was inspired by, such as those of the architect or the game designer (although a case could be made that some works in those fields are derivative, but that is another discussion). I will leave it to others to develop approaches from the ingredients presented in this book that focus more on this process of creating entirely new organizational arrangements, which I am sure is possible. It would probably entail giving the organizational designer a more leading role. As I discussed earlier in this chapter, the organizational designer in game based organization design is not the one shaping the organization. He or she merely directs the process and lets the participants originate all the knowledge and ideas. I could imagine another interpretation of this role that relies more on the organizational designer's capacity for abductive reasoning. The problem I see with that approach is that the result of the process will become dependent to a much larger extent of the qualities and abilities of the designer, which are usually shaped through extensive experience and to some degree through talent. So I do not foresee a future in which the Frank Gehry, the Will Wright (designer of classic games such as SimCity[40]), or the Jonathan Ive of organizational design will be on the cover of Time magazine. My aim would be to strive for a future in which managers are better equipped to understand and shape their organizational systems, without having to rely on 'star designers'.

Despite the limited amount of abductive reasoning taking place, game based organization design does leave open a possibility for designing a new organizational entity, as long as that entity is an extension of the existing organizational context of the co-designers. This was the case with the ECC project, which was an addition to an existing hospital. Working from their own experience and context, the co-designers and stakeholders in that project generated ideas and built on a shared understanding of the new center. However, I would hesitate to characterize what I observed in those workshops as abductive reasoning. There was a tendency to stay close to what was familiar. Another option that is not precluded by a lack of abductive reasoning is a reconfiguration or redesign of an existing organizational system. This is something I will talk about in Chapter 7.

6.6   Towards an application of game based organization design

If the result of game based organization design is a better understanding for all stakeholders of the organizational system they are part of, then this is a valuable outcome in itself. But this chapter was about design. And design is not just about explaining and understanding a phenomenon, but also about building on this understanding by producing a fitting course of action. A better understanding of an organizational system can form the starting point for a change to that system. I will talk about making changes to an organizational system in the next chapter, because it is closely related to the subject of strategy.

Theoretically speaking, the understanding, testing, and adaptation of organizational rules could lead to a cyclical process of constant improvement. This would be in line with the action orientation that is advocated in the 'managing as designing' perspective. In a 2008 article in a special issue of *Organization Studies* on that subject, Raghu Garud and his co-authors talk of a pragmatic approach to organizational design that views design as continually evolving, given the ever-changing environment that the organization has to deal with.

> Eventually, a pragmatic approach involves the fusing together of two meanings of design, that is, as both process and outcome. Any outcome is but an intermediate step in an ongoing journey, representing both the completion of a process as well as its beginning. Whereas the scientific approach emphasizes the need to crystallize designs, the pragmatic approach highlights the value of retaining fluidity.[41]

I am quite partial to this view of design as an ongoing journey and of the outcome as being just an intermediate step. I believe it is conceptually elegant and it fits the way modern organizational arrangements such as the open source and Wikipedia communities are shaped (both of which were used as exemplary cases by Garud). It also fits closely to my findings in applying the video game design process to organizations. Whereas I started out with the aim of a concrete end result for the design process – in my view, this was to be the rule set – I encountered reactions that instead saw this rule set as a stepping stone to further design efforts (ECC) or as 'a globe to travel across' (WBTM). And yet, I have to acknowledge that this conceptual elegance of the ongoing journey may have a limited bearing on the managerial relevance of the approach. In my experience, it is generally hard to sell a journey to a manager. Business leaders are interested in results, not in travels. So in the coming, final chapter, while attempting to maintain a certain conceptual elegance, I will expand the discussion of game based organization design to concrete results, placed in the context of strategy. To close off the current chapter, I want to present an example of one the new artifacts I introduced to organizational design, the paper prototype, and I will compare and contrast the process I have described in the previous sections to an approach that has been around for much longer, simulation gaming.

6.7   An example of a paper prototype: The IT Community

To complement the rather abstract descriptions of methods and approaches in this chapter, I will provide an example of what a paper prototype can look like. It is a game based organization design project that took place at the head office of a Dutch bank. The context for this project is a so-called IT Community. This is an online platform that the bank uses for discussions and knowledge exchange about IT related subjects. The IT department at this bank supports some 40,000 workstations, smartphones and other devices, and fields around 400,000 calls annually. One of the goals of setting up the IT Community platform is to reduce this number of calls by having employees share tips, tricks, and solutions to common problems amongst themselves. A longer-term goal would be to use the IT Community to get more input about user requirements and to have discussions about the goals and strategy of the IT department. But in order to reach these goals, the number of users of the IT Community needs to increase, and they need to become more active on the platform. The main objective of the project in question was to design a strategy for further activating the IT Community, in the form of concrete measures or interventions that the IT department could undertake.

I will not discuss the entire project here, but limit myself to describing the paper prototype. The description is of the second iteration of that prototype, after adjustments were made based on a first round of playtesting. The prototype takes the form of a board game. See Figure 6.5 for some of the play materials.

In the game, two interconnected processes take place. Players of the game are divided into two groups, the managers and the participants (of the IT Community). The managers take measures that aim to increase the number of participants or increase their activity. The participants use the IT Community to ask and answer questions, and give and receive information.

Managers can score points by implementing measures. For this, they need to select one of the measures they received on cards at the start of the game. These measures and the effects they have on the IT Community were a result of earlier brainstorms during the design process. To implement a measure, players need to collaborate with one of the other managers. They need to agree on who will carry which part of the cost of the measure. If they reach an agreement, the measure is implemented, each manager receives a number of coins, and the specified number of participants is added to the IT Community (in the form of pieces that are placed on the game board) or existing participants are activated (which means that pieces are moved up to a higher level).

The other group, the participants, can score points by undertaking activities in the community. The underlying idea is that there are a number of questions (printed on cards) that need to be answered. All the information that is necessary for answering these questions is present inside the community. The cards containing this information have been handed out randomly to the participants. By combining the right pieces of information, the answer to each question can be deduced. To make sure that only the information in the game is used and not the players' own knowledge, the questions are not about IT

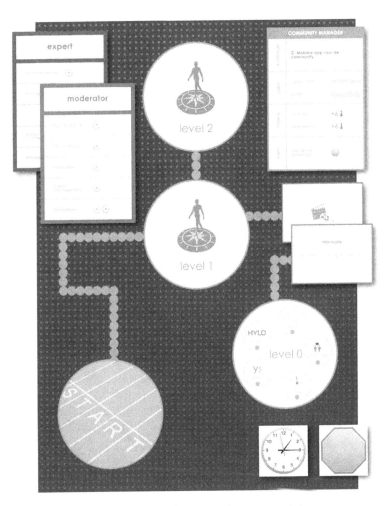

Figure 6.5   Play materials of the IT Community paper prototype

but about obscure movie trivia. Participants can give information (in the form of placing a card containing information on the board). They can collect information (by taking an information card off the board). They can ask a question (by placing a question card on the board), or they can answer a question (that was placed on the board).

The effect of these actions is that one or more pieces on the game board take the color of the player in question.

The winner of the game is the player with the highest individual score. Managers score points by implementing measures. Participants score points by undertaking activities. However, if less than twenty pieces have been brought to level two of the game board, then the activity of the community has not been raised sufficiently and there is no winner at all.

There are many more nuances to this game that I will not go into here. As I said before in this chapter, explaining a set of rules usually leads to confusion. It is by playing the game that understanding emerges.

6.8   Comparison to simulation gaming

To be able to gauge the value of game based organization design, it is necessary to compare the approach to other, similar organizational interventions. A prime candidate for this comparison is the class of interventions according to the simulation gaming approach, as discussed in Chapter 3. In this section, I will confront game based organization design with simulation gaming to uncover the main differences and similarities.[42] Before I do that, let me say a few words about the design process for simulation games.

Some authors in the simulation gaming tradition have contended that the design process of a simulation game is as important as playing the final product. It is considered by some to be 'a process of interactive and systematic strategy development'.[43] One of the founding fathers of the simulation gaming tradition, Richard Duke, described a simulation game design approach that is still the mold by which many simulation game designers work. The process not only includes design, but also construction and implementation of the simulation game. It consists of nine basic steps[44]:

1. Develop written specifications for game design.
2. Develop a comprehensive schematic representation of the problem.
3. Select components of the problem to be gamed.
4. Plan the game with the Systems Component/Gaming Element Matrix.
5. Describe the content of each cell of this matrix in writing.
6. Search the 'repertoire of games' for ideas to represent each cell.

7. Build the game.
8. Evaluate the game against the specifications.
9. Test the game in the field, and modify.

This process shares some elements with the video game design process as described in this chapter, notably its iterative approach and its use of what video game designers would call a paper prototype that is evaluated in playtesting sessions. Yet there is an important difference in the intended end result of the process. The final product of the process described above is a finished simulation game, ready to be used in the field. The result of the video game design process as described in this chapter is a set of rules that has been evaluated against the gameplay experience it was supposed to elicit.

Taking the above design process into account, I will discuss the differences and similarities between simulation gaming and game based organization design according to three themes:

1. Prototype or finished game.
2. Relation between game reality and organizational reality.
3. Extent and type of involvement by the client.

6.8.1 Prototype or finished game

In the case of game based organization design, the game is not an end product. The game is a means to an end. It is an unfinished prototype meant to express the rules and thus form a conduit for understanding and improving an organizational system. This aspect forms a major differentiator between game based organization design and the simulation gaming approach. The vast majority of simulation gaming interventions have one or more carefully moderated play sessions – and the associated learning experiences   as their result. The game that is played in those sessions is presented as a finished product, not as a prototype to be tested.

6.8.2 Relation between game reality and organizational reality

If we compare the games that are played in the final playtesting sessions of game based organization design to a typical simulation game, we can state that the former is not as prolonged, that its rules are less complex, and that it is played at a higher level of abstraction. The games of game based organization design have a direct relation to the organizational reality, but not at the level of tasks that participants have in their daily work. Many simulation games *do* get down to the level of work activities such as handling incoming phone calls, producing products, and setting prices. The games of game based organization design zoom out a bit more and allow participants to hover over the organization and its environment, as it were, each taking the role of strategist from their particular perspective (that is, the stakeholder they are representing). This also means that the games of game based organization design do not attempt to construct a detailed simulation of reality. Rather they aim to let the complexity emerge out of a limited set of rules, with the help of the players. The games in these projects are explained and played in a session of several hours, with a full round of play taking forty-five minutes to one hour. A typical simulation gaming session on the other hand will last for a full day or even several days. In terms of role-play, participants in a simulation game are often instructed to play themselves. In the games of game based organization design, participants are invited to play the role of one of the stakeholders. In some cases they represent their own group (a surgeon playing 'the surgeon') but at times they are also asked to represent other groups.

6.8.3 Extent and type of involvement by the client

The method for 'charging' the game – that is, for giving it the characteristics of the situation in question – and for involving the stakeholders form a final area of differentiation. In game based organization design, the activity of charging the game is an integral part of the design process itself. Several steps in the approach have the aim of extracting as much information as possible from the co-designers and other stakeholders, which can then be used to design the game that will be played in the final playtesting session. Extracting this information is done by creating lusory spaces (see Chapter 4). Without the participants fully realizing it, they are co-designing their own game, which they will subsequently play. The game does not take shape until the organizational designer starts building the paper prototype, although a framework is set up beforehand. This means that the final playtesting session is the first time the client team – or co-designers, as we called them earlier in this chapter – are confronted with the finished game. They are more or less on equal footing with the rest of the stakeholders and are participating in the playtesting session without being fully responsible for it. In the typical simulation gaming approach, the game is specified to a much larger extent in the first steps of the design process, as outlined earlier in this section. The client knows what he or she is going to get and can test prototype versions against those specifications. As such, there is a core client team that knows more about the aim and workings of the game than many of the other players. In fact, in most cases the majority of the participants playing the simulation game will not have been involved in its design. They are playing someone else's game.

The way in which the participants are invited in game based organization design is different from a typical simulation game as well. In game based organization design, the stakeholders are invited by the co-designers based on very limited information about what the workshops will entail. This is one of the ground rules for creating a lusory space, as discussed in Chapter 4. Furthermore, the stakeholders that are invited are often mostly external to the organization. In a typical simulation game, there is a more detailed invitation and briefing for participants. Sometimes, there is even a more or less mandatory attendance. The involvement of external stakeholders is rare. Frequently, the outside world is represented by the way the game system reacts to actions of the players or the moderators will represent external stakeholders.

Table 6.3 summarizes the most important differences between simulation gaming and game based organization design.

Table 6.3   Simulation gaming versus game based organization design

Simulation gaming	Game based organization design
Game is a finished product	Game is a paper prototype, to be tested
Game is a complex and detailed representation of organizational processes	Game is an abstract representation of the interactions between different stakeholders; complexity arises from the actual behavior during play
Play session can last several days	Play session lasts several hours
Players are instructed to play themselves	Players are invited to play the role of one of the stakeholders
Most players have not been involved in designing and/or charging the game	Most players have been involved in designing the game
Game is specified at the beginning of the design process	Game takes shape in the course of the design process
Core client team knows what the game will look like	Core client team does not know the game beforehand and is on equal footing with the rest of the players

7
Strategy

Saskia had been rather quiet. 'What did you think of your role in the game?,' I asked her. We were evaluating a playtesting session at VGZ, a Dutch health insurance company. The challenge at hand was promoting a healthier lifestyle for their clients and Saskia was representing the role of the caregiver, the friend or family member who supports a person who is trying to adopt a healthier lifestyle. 'To be honest', Saskia said, 'I thought my role was a little odd.' 'How so?' I asked. 'Well, all the other players had money to spend on measures that promoted a healthier lifestyle. But I only had all of these time coins. So everyone kept coming to me when they were running short on time, because I had more than enough of it.' The group fell silent for a moment, as they processed her comments. Then Ron, who represented the municipality in this session, said: 'Isn't that exactly what happens in our healthcare system? That we collectively lean on the caregivers, people in the client's social environment, to fill up the gaps when the money runs out?' There were nods as Leonie, the client lead for this project, said: 'That could be a real risk factor. If the caregiver collapses, so does the whole house of cards.' A weakness in the system had been identified and the group was now ready to start discussing a strategy for dealing with it.

7.1 What we talk about when we talk about strategy

I have mentioned the word strategy earlier in this book, in Chapter 5. There, I talked about rules of thumb, or heuristics, that can be considered a strategy. This can be a strategy for winning a game, but also a strategy for achieving an organization's objectives. Let me now elaborate on that and talk about how strategy can be positioned in relation to game based organization design, the approach introduced in the previous chapter.

Business historian Alfred DuPont Chandler formulated a definition of strategy after studying large-scale enterprises such as Sears and General Motors in the early 1960s. His definition is still widely accepted, so it seems like a good starting point here. According to Chandler, strategy is

the determination of the basic long-range goals and objectives of an enterprise, and the adoption of courses of action and the allocation of resources necessary for carrying out these goals.[1]

This definition combines two questions that need to be answered in the strategy process. The first is the question of 'what?' What are we trying to achieve as an organization? What business will we be in? What is our mission? This is what Chandler refers to as determining goals and objectives. Some firms expend a lot of effort on this stage of the strategy process, by means of identifying trends and market opportunities, and by performing industry benchmarking, often with the help of external consultants.

The other question is one of 'how?' How will we achieve our goals? How will we compete in this market? We could also say that 'how?' is a two-part question according to Chandler. The first part is about choosing a course of action. And the second part is about equipping the organization with the means to carry out this course of action. This second part is thus about organizational design, as discussed in the previous chapter, and the implementation of that design.

What follows as a final stage is strategy execution. It is not part of Chandler's definition and I would contend it is not part of the strategy process. By no means am I saying strategy execution is not important, but it carries with it issues of performance management and alignment that merit an extensive discussion for which I do not consider this text to be the appropriate place.[2]

So based on Chandler's definition we are left with the four-step process depicted in Figure 7.1. The first step is setting goals. The second step is determining a course of action. Let us call this course of action the strategy. It is important to acknowledge that the environmental conditions that an organization faces form an important factor in determining the strategy next to the goals, which is more or less implicit in Chandler's definition. The third step is (re)designing the organization in such a way that this course of action can be carried out effectively. And the final step is implementing the design. This four-step process includes both strategy formulation and organizational design, in that order. And thus, in Chandler's terms, structure follows strategy.

The academic field of strategic management has traditionally been concerned with performance, with answering the basic question 'Why do some organizations perform better than others?'[3] In answering this question, the field has looked at both the strategies and the structures chosen by the 'winners', as well as at the attractiveness of the industry that an enterprise is in. Michael Porter's work from the 1980s on the five forces shaping industry competition and the three generic

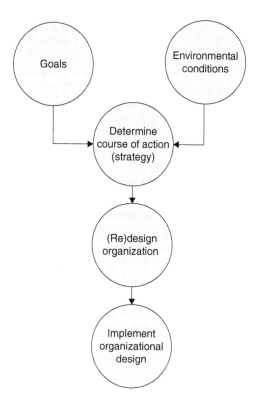

Figure 7.1   Strategy formulation and organizational design

strategies among which firms can choose (overall cost leadership, differentiation, or focus) is still very influential.[4] Another classic study was the one by Raymond Miles and Charles Snow,[5] who found four strategic types – which can be seen as combinations of strategy and structure – among organizations (defenders, prospectors, analyzers, and reactors) with each performing well in a certain context. For our current discussion of strategy, this performance focus is too narrow. I would like to broaden the view of strategy in two ways. The first is by not just looking at strategies for profit-seeking firms.[6] The definition and the process that I described earlier (Figure 7.1) still hold for not-for-profit and public organizations, although the focus in strategy research has been almost implicitly on private firms with shareholder value as the ultimate measure of performance. The second way in which I would like to broaden the view of strategy in this text is by not just looking at comprehensive organizational strategies but also at strategies for achieving specific goals at a lower level in the organization. Traditionally this level is called the operational level, but a strict separation between a strategic and an operational level would seem to me increasingly untenable in contemporary organizations. There is an ever greater necessity for entities lower in the organization to choose a course of action to address specific opportunities or threats that arise 'on the frontlines', albeit guided by a broader organizational framework. Earlier in this book, I mentioned the 'minimal structure',[7] Eisenhardt's 'strategy as simple rules'[8] and Weick's call for 'underspecification'[9] as examples of this tendency. In a study of four multinational firms, Patrick Regnér of the Stockholm School of Economics found that strategy-making in what he calls the periphery is more externally oriented and explorative, while the traditional, corporate strategy-making is more based on planning and formal reports.[10] This strategy-making in the periphery is what I would like to focus on here. We could for instance think of situations where an IT department needs to improve its collaboration with internal clients and external suppliers, or a business unit wants to be more responsive to customer queries. These types of strategies have a slightly different starting point from the one in Figure 7.1. In addition to environmental conditions, there is usually a higher-level organizational strategy in place that forms a framework inside which this specific goal needs to be addressed. In some cases the process may be reversed, when these lower-level strategies help to ultimately shape the corporate strategy.[11]

The latter phenomenon relates closely to a final dimension of strategy that is relevant here, the distinction made by Henry Mintzberg between intended and emergent strategies.[12] Not all organizational strategies are planned. It is also possible to attribute the label strategy retrospectively to a certain emergent pattern of action.

In the context of this book I am not interested in what high-performing competitive strategies look like. As I stated in the previous chapter, I am interested in equipping managers with tools. Specifically, I aim to equip them with tools for strategy formulation and organizational design. And as I explained above, I look at strategy in a broad sense. That broad sense includes both profit-seeking firms and not-for-profit organizations, and it includes both comprehensive organizational or corporate strategies as well as strategies for addressing goals or challenges at a lower level in the organization. The tools and methods of strategy-making have become the focus of a recent stream of studies under the label Strategy-as-Practice or Strategizing Activities and Practice (both abbreviated as SAP) that closely fits the perspective on strategy that I am exploring here, not only with its focus on the methods of strategy-making but also with its move away from performance and its widening of the types of organizations studied.[13] I would like to discuss game based organization design – as introduced in the previous chapter – in that light.

7.2 Redesigning an organizational system

Let me return to a number of concepts introduced earlier in this book. I have talked about the organization as a complex system and about the need for new ways of describing and designing these complex organizational systems. I also identified some similarities between organizational systems and games. Both are characterized by a complex possibility space that cannot be captured in a mathematical model and both are prone to design flaws that can lead to undesirable dynamics. I then extended the circumscriptive rules of games and the process by which they are designed to organizations. I described this game based organization design as a process that builds awareness of the mechanics of an organizational system, articulates them as rules, then plays with those rules in the form of a paper prototype to find out their possible effects. The result is a measure of awareness and reflection for stakeholders about the constituting elements of the organizational system, their interactions and the possible outcomes of this dynamic.

After having talked about strategy in the previous section, we can now be more specific about what exactly constitutes an organizational system. I would argue that the most useful way to look at an organizational system is as a goal-driven configuration of three elements: environment, strategy, and structures and processes.[14] I would also argue that it is most useful to look at the elements in that order, as also follows from Figure 7.1. So structures and processes follow strategy, which follows goals and environmental conditions. In game based organization design this distinction between the three elements of an organizational system is not made explicitly, nor does making the distinction have an added value in the approach. The added value of the approach lies chiefly in participants increasing the holistic understanding of the organizational system they are part of and exploring how the goals stated at the outset (comparable to the first step in the strategy process, Figure 7.1) can be achieved inside this system. A holistic understanding is not served by a decomposition of the system into elements. However, after having gone through the steps of game based organization design (see Table 6.2) the increased understanding with stakeholders can let a need arise to 'fix the system'. At that point, it is helpful to understand what the constituting elements of the system are. Which are the levers we can pull to change the outcomes? Those levers are environment, strategy, and structures and processes. I will assume here that the goals remain fixed. But of course a result of the process could also be that the goals are adjusted because they are deemed unrealistic after the characteristics of the organizational system have been explored. Assuming the goals stay the same, a re-design of the system would thus entail re-designing one or more of these three elements.

1. *Environment.* Re-designing the environment of an organization is obviously challenging, but not entirely impossible. Jeffry Pfeffer and Gerald Salancik were among the first to acknowledge that an organization has other options than to react defensively to its environment by adapting its strategy and structure.[15] It can also react offensively by avoiding external control, integrating parts of its environment by means of mergers, acquisitions, or joint ventures, or by lobbying its legislators.[16] One of the main advantages of game

based organization design is that external stakeholders are invited to become part of the process. If a goal can be formulated that unites all stakeholders and a shared understanding of the organizational system can be created, then this can form a sound basis for taking joint actions that re-design the environment in the desired fashion.

2. *Strategy.* Generating courses of action for achieving the stated goal or goals is an integral part of game based organization design. In the approach, these are the lower-level strategies I talked about in the previous section, called core mechanisms in game based organization design. All stakeholders contribute to producing these strategies. After going through the steps of the approach (Table 6.2) and having experienced the organizational system in action, it is possible to have the stakeholders select the courses of action they deem most effective. It is also possible to construct a more comprehensive organizational strategy with these lower-level strategies as the ingredients.

3. *Structures and processes.* This element is what we referred to in the previous chapter as the end result of the organizational design process. That means structures and processes should be understood here as being underspecified. The rules that underlie the paper prototype can form a good framework for the re-design of the structures and processes of the organization. Once there is agreement on the paper prototype among the stakeholders, the circumscriptive rules that are at the core of the prototype can be used as a framework inside which the design of business processes and organizational structures can be detailed. However, I should stress that these rules can lose much of their value when put into writing. Preferably, those charged with these follow-up design efforts should have playtested the paper prototype themselves to fully understand the rules and their dynamics. Only then can a documentation of the rules have its full value as a framework.

A structured discussion after the playtesting session (the final step in game based organization design, see Table 6.2) can be a useful jumping off point for the re-design of one or more of these three elements of the organizational system. The central issue in this discussion should be determining the extent to which there is a misfit between the goals and the organizational system. Based on this discussion, the organization can decide to conclude the game based organization design process or to continue the process by going into a re-design cycle. Concluding the approach can either be because the increased awareness with stakeholders about the organizational system is considered sufficient result in itself, or because the results are used as a starting point for a re-design effort by traditional methods. What I mean by traditional methods is activities such as improving business processes, preparing new job descriptions, or having a plan drawn up and approved that contains a new strategic direction. I will not go into these types of approaches here, since there is more than enough literature available on these subjects.[17]

Alternatively, the organization can decide to continue their re-design efforts using game based organization design. This would entail a re-design of the paper prototype by the organizational designer with the help of the co-designers, to reflect the proposed changes to environment, strategy, structures and processes. Then, an additional play-testing session should be convened to test the new design. This could lead to agreement on the new design or to another cycle of adaptation of the paper prototype. Once there is agreement on the design of the organizational system, what remains is implementation, which is a subject I will say a few words about in the next section.

By including this redesign effort, game based organization design could cover all the steps of strategy formulation and organizational design outlined in Figure 7.1, save for implementing the design. It should be noted that the step of setting goals in Figure 7.1 is only dealt with superficially in game based organization design. It is by no means the extensive – and often expensive – exercise normally associated with the strategic positioning of an organization. The approach assumes this positioning has already taken place and only asks of the co-designers to articulate the goal or goals that will be starting point of this design effort.

Figure 7.2 shows how game based organization design can be extended with this re-design effort. The size of the circles indicates whether the activity is done by just the organizational designer (smallest circle), with the help of the co-designers, or together with all the stakeholders (biggest circle). I should stress that this extension of game based organization design has not been widely applied or rigorously tested in the field. To a certain extent, I am making a theoretical extrapolation here. But later on in this chapter, I will give an example of a project that comes close to the approach depicted in the figure below.

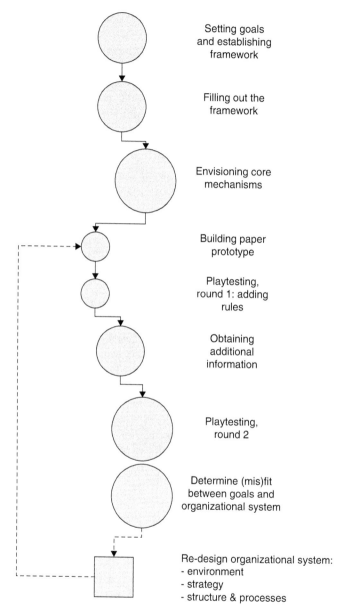

Figure 7.2 The extended game based organization design approach

7.3   Implementing the design

It is important to note that the process depicted in Figure 7.2 is a design process. That is to say, the end result of the process is a description, not an implementation. What remains to be done is implementing the redesign of the organizational system and executing the strategy. This is by no means a trivial matter. In fact, some authors claim that effective implementation and execution are what separate the unsuccessful from the successful organizations.[18] But it is a subject that is largely outside the scope of this text. I can refer the reader to the many excellent titles that are available about strategic alignment, change management, and project management.[19] There are just a few remarks I will make about these subjects here. By its action-orientation and its inclusion of stakeholders, game based organization design lays a sound foundation for implementation and change management work. There are two elements to this foundation.

The first element is that game based organization design includes an evaluation of the redesign (through the use of playtesting) as part of the process. This diminishes the chance of surprises coming to light once what has been designed is implemented, because the effects of the (redesigned) organizational system have already been explored. We can view this as an important lead-in to implementation. Still, despite this action-orientation, there comes a point in the process – somewhere at the boundary of design and implementation – when what has been designed should be documented. Game based organization design (as depicted in Figure 7.2) can result in two types of written artifacts that can be considered boundary objects[20] between design and implementation:

- *A set of circumscriptive rules.* These can be extracted from the paper prototype and can form the basis for implementing the re-design of the environment (such as new types of collaborations between stakeholders) and the re-design of structures and processes (such as changes to information systems or the workplace). As mentioned before in this and previous chapters, those who undertake follow-up activities with this set of rules should have been involved in a playtesting session of the paper prototype. Otherwise, a full understanding of them is all but impossible.

- *The outline of a strategy.* As part of game based organization design, courses of action (called core mechanisms in the approach) are generated, selected, tested, and adjusted. The courses of action that survive this process can be documented with the aim of implementing them (as was done in the project I will describe later on in this chapter). Alternatively, the courses of action can be integrated and documented in the form of a higher-level organizational strategy.

The second aspect of the foundation for implementation efforts that game based organization design lays is the inclusion of all the stakeholders in the process. These stakeholders have been involved in the re-design effort and have experienced the re-designed organizational system in action. This will likely increase their buy-in to the design that is the result of the process and should form a sound starting point for broader change management efforts. These

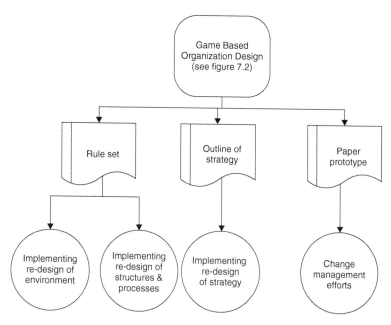

Figure 7.3 The boundary objects between game based organization design and implementation

broader change management efforts could include sessions in which other groups of stakeholders play the paper prototype as a way to increase their awareness of the changes that the re-design of the organizational system entails. That would make the paper prototype a change management instrument rather than just a conduit for the design efforts, and this could mean a change in the requirements the prototype needs to meet (or even doing away with the term 'prototype'). This use of the paper prototype as a tool for change management is pure conjecture at this point, but an interesting subject for further research. It would correspond with the use of games in the simulation gaming tradition, as discussed earlier in this book (in Chapters 4 and 6).

Figure 7.3 shows the artifacts that can come out of game based organization design and their relation to implementation and change management efforts.

7.4   The added value of game based organization design

With the extension to strategy in this chapter, I have concluded my argument for a game based approach to organizational design. In closing, let me address where I believe its added value lies. In the most general sense, I believe game based organization design is one way to address the gap between conventional tools for organizational design and strategy formulation on the one hand and the practical strategy and design work needed in contemporary organizations on the other. A study by Johanna Moisander of the Helsinki School of Economics and Sari Stenfors of Stanford University shows exactly this gap between the available tools, which are based on a paradigm of rational problem solving by individual decision-makers, and the tools that are called for, which 'support collective knowledge production, promote dialogue and trust, and function as learning tools'.[21] This describes the essence of the gap that game based organization design is attempting to bridge.

There are two additional elements of added value that I have tried to get across in this book. The first is the action-orientation of the game based approach. Participants in the process are enabled to actively explore their organizational system and to ultimately play with that system in the form of a paper prototype. This makes it possible to evaluate the choices that are made in the course of the design process by 'living through them' and it likely increases the support for these choices. The action-orientation forms a contrast with traditional methods of strategy formulation and organizational design that have reports, meetings, and PowerPoint presentations as their forms of expression. The playtesting technique that is central in this action-orientation of game based organization design also plays a role in preventing unintended consequences of design decisions. Possibilities for gaming the system (discussed in Chapter 5) are uncovered in the course of the design process, instead of after the design has been implemented.

To illustrate this need for action-orientation, let me elaborate a bit more on the final round of playtesting in game based organization design. In this playtesting session, the paper prototype board game is played by all the stakeholders. The prototype expresses the organizational rules that have been uncovered during the design process. At the start of the final playtesting session at WBTM (a project described in Chapter 5), participants needed some time to grasp the rules of the game. There were quite a few frowns after the initial explanation of the rules and the trial round was rather laborious because of the large number of actions and interactions needed to complete one turn. As one participant said:

> You [as the moderator] said very convincingly: 'It is actually a very simple game.' Well, that was not the first impression I got of it.

But through playing, there was a fast progressive understanding of the rule set and the game became more intense (Figure 7.4). There is a noticeable contrast between an apparently impenetrable set of rules when put into words or writing and an immersive experience once these rules are put into action, which points to the importance of bringing designing from the immaterial domain of thinking to the material domain of acting.[22] It is doubtful that a less hands-on approach for the evaluation would have produced an equally deep understanding with the participants of the mechanics of the design and the dynamics of the group collaboration involved in it. The activity of playtesting offers a manner of evaluation in the course of the design process which is not yet common in organizational design practice. It will come as no surprise that playtesting is considered one of the most important activities in the video game design process.[23]

For example, Microsoft Games User Research conducted over 3,000 hours of playtesting with more than 600 players for the video game Halo 3.[24] What my use of playtesting in the context of organizational design suggests is that several rounds will be necessary for it to be most effective, preferably with different groups of players. If organizational resources are limited, it is sometimes possible to combine several rounds of playtesting in one session. This can be done by making adjustments to the prototype after one round and playing the next round with the adjusted version, during the same session.

The second element of added value that I would like to highlight once more is the fact that this approach explicitly invites external stakeholders to become part of the design process and thus makes the organization's environment an integral part of the system under design. As I have tried to show at the outset of this book, this is an indispensable part of an effective approach for organizational design. Contemporary organizations have permeable boundaries and are part of increasingly complex systems. Instead of trying to isolate themselves from this complexity or attempting to reduce it, they should

Figure 7.4   A playtesting session at WBTM

find ways to embrace it. I hope that I have been able to show that game based approaches can be a way forward for business leaders and organizational scholars faced with the challenge of describing, explaining, and reconfiguring complex organizational systems.

7.5   An example of game based organization design: VGZ health insurance

One of the big challenges in public health is encouraging a healthy lifestyle. Not only is it challenging because it pertains to changing a person's behavior, but perhaps even more so because there are so many stakeholders involved. These stakeholders have many different interests and do not always speak the same language. There are stakeholders such as the different departments of the health insurance company itself, the employer, the municipality, and the physician. Adding to the complexity is the fact that responsibilities in the area of prevention and encouraging a healthy lifestyle are divided among those stakeholders. VGZ, a Dutch health insurance company, decided to take the lead in this field and resolved to come up with new strategies to encourage a healthier lifestyle, together with all relevant stakeholders. I was asked to direct a project that used game based organization design (Figure 7.2) to achieve this aim.

The first step in the process was to assemble a team of co-designers. We brought together two innovation managers and a marketing manager in this core team at VGZ. I conducted interviews with these co-designers individually to get a better understanding of the 'playing field'. How would they describe the goals that we were aiming for with this project? What was needed to achieve those goals? Which stakeholders played a role in realizing these goals? The interviews supplied me with the information needed to draw up a framework diagram that summarized the most important elements in the system at hand (Figure 7.5).

The next step was filling out this framework. With this aim, a workshop with the co-designers was convened, which I moderated together with a colleague. In this workshop we used lusory spaces (see Chapter 4) to force the co-designers to think about the elements of the framework. What are the different types of healthy behavior? What are the results of healthy behavior or the costs of unhealthy behavior? Which reasons are there not to have a healthy lifestyle?

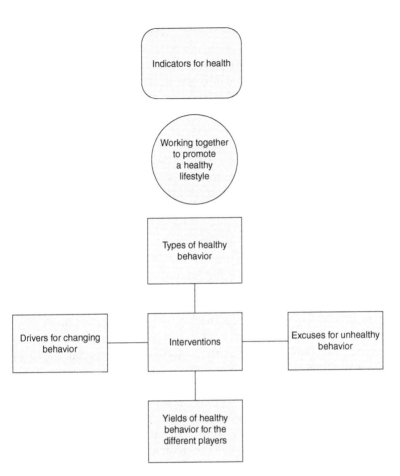

Figure 7.5   Framework diagram for promoting a healthier lifestyle

(They came up with sixty of these excuses) An important output of this first workshop was the list of the seven most relevant stakeholders, or players (to use the game terminology).

Representatives of these stakeholders were then invited to a second workshop. Gathered around the table in that session were several departments of VGZ, representatives of a municipality, a healthcare provider, and an employer. The goal of this second workshop was to envision 'core mechanisms'. What can the different stakeholders do to achieve the goal of promoting a healthier lifestyle? Devising these core mechanisms was done in several rounds of 'speed-dating', in which stakeholders brainstormed in pairs about courses of action they could take together to contribute to the overarching goal. The core team of co-designers had opened the workshop with a presentation about this goal – or epic meaning, in game terms – of promoting a healthier lifestyle. One of the elements of this second workshop was that the ideas generated during the speed-dating sessions could be 'neutralized' by the player who represented the person whose lifestyle we were trying to improve. He could use one of his excuses (the ones the co-designers had come up with during the first workshop) to do this. At the end of the second workshop, fifteen courses of action – or interventions, as VGZ likes to call them – to promote a healthier lifestyle remained. One example was 'handle with care teams' at sports clubs for people that want to do sports but need special attention. Another example was organizing a 'healthy living week' at schools. The client lead expressed her amazement and satisfaction that this relatively simple method had led to such a volume of creative ideas. Another important result of the day was, according to her, that the different stakeholders had gotten to know each other and had been able to express their interests and expectations in a safe, playful setting.

All the information that had been gathered during the first two workshops formed the input for building a paper prototype. I took this role of organizational designer upon me together with my colleague. The aim was to build a paper prototype that expressed the dynamics of this playing field. Filling out the framework diagram (Figure 7.5) formed the basis for this part of the process. Through several rounds of building, playtesting, and adapting the (pencil-and-paper) game, we arrived at a first viable iteration of the paper prototype. We then consulted our co-designers at VGZ to obtain the additional information that was needed to complete the game, such as the costs of specific interventions. After this, we were able to produce a version of the paper prototype that could be played independently by our players and that looked fairly polished (without looking completely finished).

All the stakeholders were then assembled for a second time to participate in a playtesting session. They encountered a board game that was constructed from the puzzle pieces that they, along with the co-designers, had handed to us in the previous workshops. In the game, players had to cooperate to implement interventions that raised one or more of the indicators for a healthy lifestyle of our target population. The goal of the game was to reach a certain target value for these indicators. Every intervention that was available in the game – they were the result of the speed-dating sessions in the previous workshop – had a specific contribution to these indicators but also contributed to the stakeholder's individual goals. Negotiations had to take place among the stakeholders and with the client in question – this role of client rotated among the players – about the amount of time and energy he or she would invest in this intervention.

After playing several rounds of the game, the group had a discussion about their experiences during this playtesting session. They talked about changes to the paper prototype that would make it more closely resemble the actual healthcare system. One of those changes was that not implementing an intervention – because the stakeholders could not come to an agreement – would have to lead to a decline (unhealthy behavior would increase) instead of a continuation of the status quo, which was the case in the current game. So the game also had to be about minimizing the loss, instead of just maximizing the gains. Besides talking about the changes to the paper prototype, the group also had a productive discussion about new insights they had obtained from playtesting this game. These insights were mainly to do with how collaborations between the stakeholders took shape on this playing field. Some stakeholders, such as the municipality, realized they would have to collaborate much more with others to achieve this shared goal of promoting a healthier lifestyle. They had been used to more solitary strategies. It also became clear that the recipient of these planned interventions for a healthier lifestyle played different roles. She could be an employee, a citizen, a patient or a client, depending on which stakeholder would approach her. She would react differently to a proposal from her employer than to a proposal from her physician.

After this playtesting session, VGZ decided to take the project one step further. An additional workshop was organized together with a municipality that VGZ intended to collaborate with. We played the game – which had been refined based on the first playtesting session – with a group that had not been involved in the design process. Through playing the game, we aimed to reach a shared understanding of the context. After a few rounds of play, we then basically went through another iteration of the design process. We asked participants to come up with improvements to the game, and thus to the organizational system. We also repeated the speed-dating sessions we had done earlier on in the process with these different representatives of the same stakeholders. At the end of this workshop, two interventions for promoting a healthier lifestyle had been selected by the group as being the most effective in this context. These interventions were further elaborated and implemented in the months following the session. As the client lead looked back on the process, she said:

> In the end it turns out that these playtesting sessions function as a kind of pressure cooker. In one afternoon, you can lead a group through an entire orientation process of getting to know each other and each other's positions. The session produces a number of concrete ideas that have the commitment of all the stakeholders.

7.6 Future work

In this book, I have argued for a new way of looking at organizational design and strategy formulation. I identified a gap between the current organizational reality and the tools and methods available. I have tried to show that one of the ways to bridge this gap is to introduce insights and approaches from game design into the design of organizational systems. In this chapter and the previous one, I have described a specific approach to do this, called game based organization design. Let me stress that this is but one way to put into practice the ideas I put forward in this book. I would urge readers to pay equal attention to the broader notions presented here in addition to the specifics of the approach I describe. The game based organization design approach is a first implementation of these ideas, which has proven successful in a number of projects. But I expect that there are many other ways to apply video game design in the context of organizational design and strategy formulation and I look forward to seeing this future work. I strongly believe that the challenges that organizational leaders are facing today call on their courage to let go of traditional approaches. There is unmistakably a need for new models, and I have tried to show that games, play, and game design can form a valuable source for this. But as I stated, I think there is much more to be discovered at this intersection of games and organizations.

With regards specifically to game based organization design, one of the most obvious future projects – which is already underway – would be to more broadly employ and evaluate this approach. Comparing findings from different organizational systems, by different organizational designers and for different design goals would be a step towards refining the method and more clearly identifying appropriate contexts, which will likely increase its usefulness. As I mentioned earlier in this chapter, there are two areas of application that particularly need additional empirical evaluation. One is the extension of game based organization design with a re-design cycle of the organizational system (as depicted in Figure 7.2). The second is the follow-up of game based organization design with implementation and change management activities, such as the wider use of the paper prototype.

Looking beyond game based organization design, I think we may be only at the cusp of the 'ludic age', as optimists such as game designer Eric Zimmerman have called it.[25] Video games and play in general are exerting an ever-stronger influence on our culture, which will inevitably seep through more and more to organizational life as well. I consider this book to be a contribution to this development. I have briefly touched upon the broader movement towards 'playful organizations' in Chapter 4, but there is much more to say about this subject than I can cover here. What I have described in this book is embedding ideas about play (lusory spaces) and games (paper prototyping and playtesting) in an organizational design process. The organizational design process is typically insulated from the day-to-day business of an organization, in terms of time and place. A playful organization, on the other hand, would have the ambition of breaking out of that insulation and introducing elements of play and games more widely, to make them part of the organization's culture and processes. Harald Warmelink of the Delft University of Technology has made a first effort of identifying what the elements of such a playful organizational culture would be. He mentions six values[26]:

- Contingency: employees appreciate uncertainty and eventuality.
- Agility: employees can act on any opportunity.
- Equality: employees have equal opportunities.
- Teachability: employees take opportunities for a large variety of learning experiences.
- Meritocracy: employees are socially recognized for their efforts and competence.
- Conviviality: employees interact informally.

I mentioned the attractiveness to managers of many of the things that transpire in video games, such as some of the elements identified by Warmelink, in Chapter 3. But I also talked about a number of obstacles that make realizing playful organizations a challenging endeavor (in Chapter 4), such as the fact that many of the great thinkers on the subject of play stress its careful isolation from the rest of life. But in no way does that imply that I do not consider the move towards playful organizations a promising development. I do believe that much future work remains to be done, both by organization researchers and by brave practitioners in the field.

Apart from these broader ideas and perspectives on future developments, I have presented here a concrete approach in the form of game based organization design. Although the ideas presented in this book are relatively new and game based organization design can certainly not be called mature, it is an approach that has been tested and that is being used today. I look forward to hearing your experiences in using, evaluating, adapting, and extending it. And I hope that together we can continue our work on tools for improving the organizations we live in.

Notes

1 Introduction

1. Published by Sony Online Entertainment in 1999.
2. Published by Blizzard Entertainment in 2004.
3. van Bree (2013).
4. van Bree (2013).
5. First reported in van Bree, Copier, and Gaanderse (2010).
6. First reported in van Bree and De Lat (2011).
7. Eden and Huxham (2006).
8. Eden and Huxham (2006), p. 397.
9. All interviews were conducted in Dutch. The translations used for the examples are the responsibility of the author.

2 Systems

1. As related by Stewart (2009) in his excellent account of this crucial week, based on interviews with the key players.
2. Katz and Kahn (1966); see also Chapter 4 of Scott and Davis (2007) for an overview of the open systems perspective in organization theory.
3. Castells (2000).
4. Meadows, Club of Rome et al. (1972).
5. Forrester (1968) is a good starting point for exploring System Dynamics.
6. See Mingers and White (2010).
7. Ackoff (1979) is his most well-known and vicious critique of operations research.
8. Taylor (1911).
9. Fayol (1918).
10. KPMG International (2011).
11. IBM (2010).
12. IBM (2010), p. 3.
13. Hamel (2007).
14. Peters and Waterman (1982).
15. See Lashinsky (2012) for a look inside Apple, limited as it is because of the company's secrecy.
16. For an extensive and well-informed look at Google's founding and growth, I recommend Levy (2011).
17. See Semler (2004) for more on his management philosophy.
18. Tapscott and Williams (2006).
19. See, for example, Davis and Marquis (2005) for this warning; for a general explanation of social mechanisms, see Hedström and Swedberg (1998).

20. Neumeier (2009), p. 41.
21. See, for example, Mingers and White (2010) and Gharajedaghi (2011).

3   Games

1. Rovio (2011).
2. Rovio (2013).
3. According to data by video game research firm EEDAR presented at the Game Developers Conference in San Francisco on March 28, 2013, reported in Venturebeat (2013).
4. Notably Space War, developed by Steve Russel at MIT in 1962.
5. Introduced by Taito in 1978.
6. Introduced by Namco in 1980.
7. Released by Universal in 1980.
8. Published by Brøderbund in 1983.
9. These figures are from Malliet and De Meyer (2005), which offers an excellent and detailed account of the history of video games, including its controversies as to who should be considered its founding father.
10. Released by Eidos Interactive in 1996.
11. Based on data by market research firm NPD Group (www.npd.com).
12. Edwards (2011).
13. Graft (2013).
14. Newzoo (2012).
15. Based on Newzoo (2012) for the US and other 2012 data by Newzoo (www.newzoo.com).
16. Some further reading in this regard: Gee (2003) introduces video games as an instructional tool, an idea further developed by Shaffer (2006). Beck and Wade (2006) paint a fairly optimistic picture of the influence that the 'gamer generation' will have on the workplace, an idea also touched upon by Seely Brown and Thomas (2008). Game designer Raph Koster (2005) makes a convincing case for learning as the core of why good games are fun to play.
17. Figures are based on industry statistics by the NPD Group (www.npd.com).
18. This genre of games is also referred to as Massively Multiplayer Online Role-playing Games, or MMORPG's. I will use the shorter acronym MMOG throughout this text, foregoing the emphasis on the role-playing aspect that the extended abbreviation adds.
19. Turkle (1995).
20. Turkle (1995), p. 14.
21. For more on these spillover effects and the fading boundary between online games and daily realities, I recommend Copier (2007) and Taylor (2006).
22. Cherney (1999).
23. Morningstar and Farmer (1991) contains their account of setting up Habitat.
24. Released by Sony Online Entertainment in 1999.

25. Rose (2012).
26. The figures can be found in Schiano et al. (2011).
27. Castronova (2005), p. 19.
28. Dibbell (2006).
29. This is a game design pattern that Björk and Holopainen (2005) term privileged abilities: 'abilities that let agents perform actions not readily available to others.'
30. The role of quests in fostering collaboration and community building has been noted by authors such as Wolf (2007) and Dickey (2007).
31. Williamset al. (2006).
32. Chen, Sun, and Hsieh (2008).
33. As noted by Fairfield and Castronova (2006), among others.
34. Grudin (1994).
35. McGrath and Hollingshead (1994).
36. Daft and Lengel (1986), p. 560.
37. See Schmitz and Fulk (1991).
38. Walther (1996).
39. Schmidt (2002), p. 286.
40. Dourish and Bly (1992).
41. Schmidt (2002), p. 290.
42. Harper, Hughes, and Shapiro (1989).
43. Heath and Luff (1992).
44. Tee, Greenberg, and Gutwin (2006).
45. Gutwin, Penner, and Schneider (2004).
46. Kraut, Fish, Root, and Chalfonte (1990).
47. Ackerman (2000).
48. This was one of the things championed by Byron Reeves in the book he wrote together with J. Leighton Read (2009).
49. A term he introduces in 'How computer games help children learn' (2006).
50. Bennerstedt, Ivarsson, and Linderoth (2012).
51. Warmelink (2013), pp. 224–225.
52. Deterding, Khaled, Nacke, and Dixon (2011) can be credited with this first attempt at a definition.
53. Developed by Kevan Davis in 2007.
54. http://www.thefuntheory.com.
55. http://nikeplus.nike.com.
56. http://fold.it.
57. As reported in Khatib et al. (2011).
58. An early critic of gamification who took this position was Margaret Robertson (2010), in an influential article on her website entitled 'Can't play, won't play'.
59. Geelen, Keyson, Boess, and Brezet (2012).
60. Hamari (2013).
61. M2 Research (2012).
62. Ryan and Deci (2000).
63. For an overview, see Medina (2005).
64. Lepper and Malone (1987).

65. Csikszentmihalyi (1990).
66. See, for example, Steinkuehler (2004).
67. Laurel (1993).
68. Also mentioned by Steinkuehler (2004).
69. Ryan, Rigby, and Przybyblski (2006).
70. Pink (2009).
71. Koster (2005).
72. See, for example, Ryan, Rigby, and Przybylski (2006), Jakobsson and Taylor (2003), and Steinkuehler (2005).
73. McGonigal (2011).
74. Mollick and Rothbard (2012).
75. For some more background on the origins of simulation games, see Mayer (2009) and Geurts, Duke, and Vermeulen (2007).
76. More on this phase of the development of simulation games can be found in Faria, Hutchinson, Wellington, and Gold (2009).
77. See Romme (2004) for more about Microworlds.
78. Such as De Caluwé (2001).
79. See Geurts, Duke, and Vermeulen (2007) and Geurts and Vennix (1989).
80. These five goals are based in part on Stoppelenburg, De Caluwé and Geurts (2012).
81. Kolb (1984).

4 Play

1. Originally published in 1938. I will use the Dutch 1950 edition as a reference here, unless otherwise indicated.
2. This issue has been extensively discussed in the context of game studies; Zimmerman (2012) offers an interesting review.
3. The sentence is translated – incorrectly, in my opinion – as 'All play has its rules' in Huizinga (1955).
4. A point made by game theorist Jesper Juul (2005).
5. Caillois (1961).
6. Suits (1978).
7. For those who are interested in this perspective, Winnicott (1971) remains an indispensable text.
8. George (2002).
9. See Spear (2004).
10. Total Google revenue for 2012 (excluding Motorola Mobile), as reported in Google (2013).
11. Levy (2011) contains an interesting account of the origins and development of this playful culture at Google in part three.
12. Sutton-Smith (1997).
13. Sutton-Smith (1997), p. 11.
14. Sutton-Smith (1997), p. 11.
15. For more on organizational improvisation, see Kamoche, Pina e Cunha, and Vieira da Cunha (2003).

16. For a general introduction to issues with the transfer of learning, see Mestre (2005).
17. Kark (2011).
18. Kolb and Kolb (2010).
19. See, for example, Geurts, Duke, and Vermeulen (2007).
20. Mainemelis and Ronson (2006), p. 81.

5 Rules

1. For a wide-ranging discussion of cheating in video games, see Consalvo (2007).
2. See Scholten (2013).
3. As discussed in Wright (2010).
4. One of the many issues highlighted in Sahlman (2010).
5. See McLean and Elkind (2003).
6. See Cooper (2008) for a full account of the WorldCom accounting fraud.
7. McCoy (2013).
8. BBC News (2012).
9. BBC News (2012).
10. The work on which much of modern game theory is based is Von Neumann and Morgenstern (1944).
11. Poundstone (1992), p. 118.
12. Both the *tic-tac-toe* decision tree and its strategy are described from a game theoretical perspective in Poundstone (1992).
13. Based on Salen and Zimmerman (2004) and Van Mastrigt (2006).
14. Brindle (2013).
15. See Anderson (1999).
16. See, for example, Pina e Cunha and Vieira da Cunha (2006).
17. Both the empirical and the modeling approach are reviewed in Van de Ven, Ganco, and Hinings (2013).
18. Scott (2008).
19. Boisot and Sanchez (2010).
20. See Eisenhardt and Sull (2001) for a general description and Davis, Eisenhardt, and Bingham (2009) for an exploration of this principle through mathematical modeling.
21. Eisenhardt and Sull (2001), p. 111.
22. See Pina e Cunha and Vieira da Cunha (2006).
23. van Aken (2004).
24. Quinn (1980).
25. van Aken (2004), p. 230.
26. See Business Rules Group (2000).
27. Business Rules Group (2000), p. 5.
28. Based on Eisenhardt and Sull (2001), van Aken (2004), and Business Rules Group (2000).
29. Salen and Zimmerman (2004) make this important distinction on p. 121.
30. Whitson (2010).

6 Design

1. Boland and Collopy (2004a), p. 3.
2. Boland and Collopy (2004c).
3. Throughout this text I will use 'organizational design' and 'organization design' interchangeably.
4. Simon (1996) p. 113.
5. Lawrence and Lorsch (1967).
6. Galbraith (1977).
7. Nadler and Tuschman (1997).
8. Goold and Campbell (2002).
9. Mintzberg (1983).
10. For more on this shifting perspective, see Jelinek, Romme, and Boland (2008).
11. Fraser (2007).
12. Brown (2008).
13. Cross (2007), pp. 36–7.
14. Boland and Collopy (2004a), p. 4.
15. See, for example, Davis and Marquis (2005) and Dunbar and Starbuck (2006).
16. Romme (2003).
17. See Weick (2004).
18. See, for example, Hackman and Wageman (1995).
19. Gehry (2004).
20. As pointed out by van Aken (2007), among others.
21. Based on the report of Pardo's talk in Harper (2008).
22. Salen and Zimmerman (2004), p. 168.
23. Fullerton (2008).
24. Fullerton (2008).
25. Fullerton (2008), p. 211.
26. Fullerton (2008), p. 189.
27. Fullerton (2008), p. 288.
28. Salen and Zimmerman (2004), p. 34.
29. Church (2006).
30. Fullerton (2008).
31. Björk and Holopainen (2005).
32. Church (2006), p. 372.
33. 'Epic meaning' is a term used by Jane MacGonigal (2011) to describe the feeling of being part of something bigger than yourself that a large-scale online video game can give you.
34. The works mentioned in the previous section are a good start: Salen and Zimmerman (2004) and Fullerton (2008). Another useful and comprehensive book on the subject was written by game designer and educator Jesse Schell (2008).
35. Björk and Holopainen (2005).
36. Cooperrider, Whitney, and Stavros (2003).
37. Weisbord and Janoff (2010).

38. Owen (2008).
39. Gharajedaghi (2011), p. 90.
40. Released by Maxis in 1989.
41. Garud, Jain, and Tuertscher (2008), p. 367.
42. As a source of information about simulation gaming I will use Stoppelenburg, De Caluwé, and Geurts (2012), a recent and comprehensive overview of the state of the art in the field of simulation gaming.
43. Geurts, Duke, and Vermeulen (2007), p. 552.
44. Duke (1980), p. 365.

7 Strategy

1. Chandler (1962), p. 13.
2. For more on alignment and performance management, see Kaplan and Norton (2006).
3. Scott and Davis (2007), p. 310.
4. Porter (1980).
5. Miles and Snow (1978).
6. Sometimes the term policy is used for public organizations, but I would like to continue to use the term strategy in this text.
7. Pina e Cunha and Vieira da Cunha (2006).
8. Eisenhardt and Sull (2001) and Davis, Eisenhardt, and Bingham (2009).
9. Weick (2004).
10. Regnér (2003).
11. This bottom-up process is described by Burgelman and Sayles (1986); see also the influence of strategizing taking place on the middle management level as identified by Floyd and Wooldridge (2000).
12. Mintzberg (1987).
13. See Vaara and Whittington (2012) for an overview.
14. This is in line with what Van de Ven, Ganco, and Hinings (2013) call the 'configuration perspective' in contingency theory.
15. Pfeffer and Salancik (1978).
16. The agency of an organization in the face of environmental pressures is an important point of discussion in the institutionalist approach to organization theory. For the interested reader, I propose Scott (2008) as an introduction.
17. See, for example, George, Rowlands and Kastle (2004) for a practical approach to business process improvement through Lean Six Sigma, see Burton, Obel, and DeSanctis (2011) for a comprehensive overview of organizational design, and Walker (2009) for a broad introduction to strategic planning.
18. See, for example, Kaplan and Norton (2008) and Bossidy and Charan (2011).
19. See, for example, Kaplan and Norton (2006) on alignment and balanced score cards, Kotter (2012) as the authority on change management, and Newton (2007) for a practical guide for project management.

20. I am using the term as defined by Boland and Collopy (2004b), p. 268: 'an artifact [...] that serves as an intermediary in communication between two or more persons or groups who are collaborating in work'.
21. Moisander and Stenfors (2009), p. 227.
22. An appeal made by Joan van Aken (2007).
23. Fullerton (2008).
24. Thompson (2007).
25. Submarine Channel (2011).
26. See Warmelink (2013).

Bibliography

M. S. Ackerman (2000) 'The intellectual challenge of CSCW: The gap between social requirements and technical feasability', *Human-Computer Interaction*, 15, 2/3, 179–203.

R. L. Ackoff (1979) 'The future of operational research is past', *Journal of the Operational Research Society*, 30, 2, 93–104.

P. Anderson (1999) 'Complexity theory and organization science', *Organization Science*, 10, 3, 216–232.

BBC News (2012) 'Starbucks, Google and Amazon grilled over tax avoidance', *BBC News*, http://www.bbc.co.uk/news/business-20288077.

J. C. Beck & M. Wade (2006) *The kids are alright: How the gamer generation is changing the workplace* (Boston: Harvard Business School Press).

U. Bennerstedt, J. Ivarsson & J. Linderoth (2012) 'How gamers manage agression: Situating skills in collaborative computer games', *Computer-Supported Collaborative Learning*, 7, 43–61.

S. Björk & J. Holopainen (2005) *Patterns in game design* (Boston: Charles River Media).

M. Boisot & R. Sanchez (2010) 'Organization as a nexus of rules: Emergence in the evolution of systems of exchange', *Management Revue*, 21, 4, 378–405.

R. J. Boland Jr & F. Collopy (2004c) *Managing as designing* (Stanford: Stanford University Press).

R. J. Boland Jr & F. Collopy (2004a) 'Design matters for management' in R. J. Boland Jr & F. Collopy (eds) *Managing as designing* (Stanford: Stanford University Press).

R. J. Boland Jr & F. Collopy (2004b) 'Toward a design vocabulary for management' in R. J. Boland Jr & F. Collopy (eds) *Managing as designing* (Stanford: Stanford University Press).

L. Bossidy & R. Charan (2011) *Execution: The discipline of getting things done*, New ed. (London: Random House Business).

J. Brindle (2013) 'Lemons, Trawlers and a Bit of Alright', *Unwinnable*, http://www.unwinnable.com/2013/06/20/lemons-trawlers-and-a bit-of-alright/.

T. Brown (2008) 'Design thinking', *Harvard Business Review*, 86, 6, 84–92.

R. A. Burgelman & L. R. Sayles (1986) *Inside corporate innovation: Strategy, structure, and managerial skills* (New York: Free Press).

R. M. Burton, B. Obel & G. DeSanctis (2011) *Organizational design: A step-by-step approach*, 2nd ed. (Cambridge: Cambridge University Press).

Business Rules Group (2000) *Defining Business Rules: What Are They Really?* http://www.businessrulesgroup.org/first_paper/br01c0.htm.

R. Caillois (1961) *Man, play and games* (New York: The Free Press).

M. Castells (2000) *The rise of the network society*, 2nd ed. (Malden: Blackwell Publishing).

E. Castronova (2005) *Synthetic worlds: The business and culture of online games* (Chicago: The University of Chicago Press).

A. D. Chandler, Jr (1962) *Strategy and structure. Chapters in the history of the American industrial enterprise* (Cambridge: MIT Press).

C. H. Chen, C. T. Sun & J. Hsieh (2008) 'Player guild dynamics and evolution in massively multiplayer online games', *Cyber Psychology & Behavior*, 11, 3, 293–301.

L. Cherney (1999) *Conversation and community: Chat in a virtual world* (Stanford: CSLI Publications).

D. Church (2006) 'Formal abstract design tools' in K. Salen & E. Zimmerman (eds) *The game design reader: A rules of play anthology* (Cambridge: The MIT Press).

M. Consalvo (2007) *Cheating: Gaining advantage in videogames* (Cambridge: The MIT Press).

C. Cooper (2008) *Extraordinary circumstances: The journey of a corporate whistle-blower* (Hoboken: John Wiley & Sons).

D. L. Cooperrider, D. K. Whitney & J. M. Stavros (2003) *Appreciative inquiry handbook* (Bedford Heights: Lakeshore Communications).

M. Copier (2007) *Beyond the magic circle: A network perspective on role-play in online games* (Utrecht: Utrecht University).

N. Cross (2007) *Designerly ways of knowing* (Basel: Birkhauser).

M. Csikszentmihalyi (1990) *Flow: The psychology of optimal experience* (New York: Harper Perennial).

R. L. Daft & R. H. Lengel (1986) 'Organizational information requirements, media richness and structural design', *Management Science*, 32, 5, 554–571.

G. Davis & C. Marquis (2005) 'Prospects for organization theory in the early twenty-first century: Institutional fields and mechanisms', *Organization Science*, 16, 332–343.

J. P. Davis, K. M. Eisenhardt & C. B. Bingham (2009) 'Optimal structure, market dynamism, and the strategy of simple rules', *Administrative Science Quarterly*, 54, 413–452.

L. de Caluwé (2001) *Veranderen moet je leren: Over de opzet en effecten van een grootschalige cultuurinterventie met behulp van een spelsimulatie* ('s-Gravenhage: Elsevier Bedrijfsinformatie).

S. Deterding, R. Khaled, L. E. Nacke & D. Dixon (2011) *Gamification: Toward a definition, CHI 2011* (Vancouver: ACM).

J. Dibbell (2006) *Play money: Or, how I quit my day job and made millions trading virtual loot* (New York: Basic Books).

M. D. Dickey (2007) 'Game design and learning: A conjectural analysis of how massively multiplayer online role-playing games (MMORPGs) foster intrinsic motivation', *Educational Technology Research and Development*, 55, 3, 253–273.

P. Dourish & S. Bly (1992) 'Portholes: Supporting awareness in a distributed work group', *Conference proceedings of CHI'92*.

R. D. Duke (1980) 'A paradigm for game design', *Simulation & Games*, 11, 3, 364–377.

R. Dunbar & W. Starbuck (2006) 'Learning to design organizations and learning from designing them', *Organization Science*, 17, 171–178.

C. Eden & C. Huxham (2006) 'Researching organizations using action research' in S. R. Clegg, C. Hardy, T. B. Lawrence & W. R. Nord (eds) *The SAGE handbook of organization studies*, 2nd ed. (London: SAGE Publications).

C. Edwards (2011) 'Microsoft slips to No. 2 behind Nintendo in U.S. video-game console sales', *Bloomberg*, http://www.bloomberg.com/news/2011-01-13/microsoft-slips-to-no-2-in-video-game-console-sales-in-december.html.

K. M. Eisenhardt & D. N. Sull (2001) 'Strategy as simple rules', *Harvard Business Review*, 79, 1, 107–116.

J. Fairfield & E. Castronova (2006) 'Dragon kill points: A summary whitepaper', *Rational Models Seminar* (Chicago: University of Chicago).

A. J. Faria, D. Hutchinson, W. J. Wellington & S. Gold (2009) 'Developments in business gaming: A review of the past 40 years', *Simulation & Gaming*, 40, 4, 464–487.

H. Fayol (1918) *Administration industrielle et générale; prévoyance, organisation, commandement, coordination, controle* (Paris: H. Dunod et E. Pinat).

S. W. Floyd & B. Wooldridge (2000) *Building strategy from the middle: Reconceptualizing strategy process* (Thousand Oaks: Sage).

J. Forrester (1968) *Principles of systems* (Cambridge: MIT Press).

H. M. Fraser (2007) 'The practice of breakthrough strategies by design', *Journal of Business Strategy*, 28, 4, 66–74.

T. Fullerton (2008) *Game design workshop: A playcentric approach to creating innovative games* (Burlington: Morgan Kaufmann).

J. R. Galbraith (1977) *Organization design* (Reading: Addison-Wesley).

R. Garud, S. Jain & P. Tuertscher (2008) 'Incomplete by design and designing for incompleteness', *Organization Studies*, 29, 3, 351–371.

J. P. Gee (2003) *What video games have to teach us about learning and literacy* (New York: Palgrave Macmillan).

D. Geelen, D. Keyson, S. Boess & H. Brezet (2012) 'Exploring the use of a game to stimulate energy saving in households', *J. Design Research*, 10, 1/2, 102–120.

F. O. Gehry (2004) 'Reflections on designing and architectural practice' in R. J. Boland Jr. & F. Collopy (eds) *Managing as designing* (Stanford: Stanford University Press).

M. L. George (2002) *Lean six sigma: Combining six sigma quality with lean speed* (New York: McGraw Hill).

M. L. George, D. Rowlands & B. Kastle (2004) *What is lean six sigma?* (New York: McGraw-Hill).

J. L. Geurts, R. D. Duke & P. A. Vermeulen (2007) 'Policy gaming for strategy and change', *Longe Range Planning*, 40, 535–558.

J. L. Geurts & J. Vennix (1989) *Verkenningen in beleidsanalyse, theorie en praktijk van modelbouw en simulatie* (Zeist: Kerkebosch).

J. Gharajedaghi (2011) *Systems thinking: Managing chaos and complexity*, 3rd ed. (Burlington: Morgan Kaufmann).

Google (2013) *2013 Financial Tables*, http://investor.google.com/financial/tables.html, accessed 30 June 2013.

M. Goold & A. Campbell (2002) *Designing effective organizations: How to create structured networks* (San Francisco: Jossey-Bass).

K. Graft (2013) 'Wargaming kicks "pay-to-win" monetization to the curb', *Gamasutra*, http://www.gamasutra.com/view/news/193520/Wargaming_kicks_paytowin_monetization_to_the_curb.php.

J. Grudin (1994) 'Computer-supported cooperative work: History and focus', *IEEE Computer*, May 1994.

C. Gutwin, R. Penner & K. Schneider (2004) 'Group awareness in distributed software development', *Conference proceedings of CSCW'04* (Chicago).

J. Hackman & R. Wageman (1995) 'Total quality management: Empirical, conceptual and practical issues', *Administrative Science Quarterly*, 40, 309–342.

J. Hamari (2013) 'Transforming homo economicus into homo ludens: A field experiment on gamification in a utilitarian peer-to-peer trading service', *Electronic Commerce Research and Applications*, 12, 4, 236–245.

G. Hamel (2007) *The future of management* (Boston: Harvard Business School Press).

E. Harper (2008) 'GDC08: Live from Rob Pardo talks about Blizzard's approach to MMOs', *WoW Insider*, http://wow.joystiq.com/2008/02/20/gdc08-live-from-rob-pardo-talks-about-blizzards-approach-to-mm/.

R. H. Harper, J. A. Hughes & D. Z. Shapiro (1989) 'Working in harmony: An examination of computer technology in air traffic control', *ECSCW'89: Proceedings of the first European conference on computer supported cooperative work* (Gatwick).

C. C. Heath & P. Luff (1992) 'Collaboration and control: Crisis management and multimedia technology in London Underground control rooms', *Computer Supported Cooperative Work (CSCW): An International Journal*, 1, 1/2, 69–94.

P. Hedström & R. Swedberg (1998) *Social mechanisms: An analytical approach to social theory* (Cambridge: Cambridge University Press).

J. Huizinga (1950) 'Homo ludens. Proeve eener bepaling van het spel-element der cultuur' in J. Huizinga (ed.) *Verzamelde werken V* (Haarlem: H.D. Tjeenk Willink & Zoon).

J. Huizinga (1955) *Homo Ludens: A study of the play element in culture* (Boston: Beacon Press).

IBM (2010) *Capitalizing on complexity: Insights from the global chief executive officer study* (Somers: IBM Global Business Services).

M. Jakobsson & T. L. Taylor (2003) 'The "Sopranos" meets "Everquest": Social networking in massively multiplayer online games', *Proceedings of the fifth international digital arts and culture conference* (Melbourne: RMIT).

M. Jelinek, A. G. Romme & R. J. Boland (2008) 'Introduction to the special issue: Organization studies as a science for design: Creating collaborative artifacts and research', *Organization Studies*, 29, 3, 317–329.

J. Juul (2005) *Half-real: Video games between real rules and fictional worlds* (Cambridge: The MIT Press).

K. Kamoche, M. Pina e Cunha & J. Vieira da Cunha (2003) 'Towards a theory of organizational improvisation: Looking beyond the Jazz Metaphor', *Journal of Management Studies*, 40, 8, 2023–2051.

R. S. Kaplan & D. P. Norton (2006) *Alignment: Using the balanced scorecard to create corporate synergies* (Boston: Harvard Business School Press).

R. S. Kaplan & D. P. Norton (2008) *The execution premium: Linking strategy to operations for competitive advantage* (Boston: Harvard Business Press).

R. Kark (2011) 'Games managers play: Play as a form of leadership development', *Academy of Management Learning & Education*, 10, 3, 507–527.

D. Katz & R. L. Kahn (1966) *The social psychology of organizations* (New York: Wiley).

F. Khatib, et al. (2011) 'Crystal structure of a monomeric retroviral protease solved by protein folding game players', *Nature Structural & Molecular Biology*, 18, 1175–1177.

D. A. Kolb (1984) *Experiential learning: Experience as the source of learning and development* (Englewood Cliffs: Prentice Hall).

A. Y. Kolb & D. A. Kolb (2010) 'Learning to play, playing to learn: A case study of a ludic learning space', *Journal of Organizational Change Management*, 23, 1, 26–50.

R. Koster (2005) *A theory of fun for game design* (Scottsdale: Paraglyph Press).

J. P. Kotter (2012) *Leading change* (Boston: Harvard Business Review Press).

KPMG International (2011) *Confronting Complexity: Research Findings and Insights*, http://www.kpmg.com/Global/en/IssuesAndInsights/ArticlesPublications/Documents/complexity-research-report.pdf.

R. E. Kraut, R. S. Fish, R. W. Root & B. L. Chalfonte (1990) 'Informal communication in organizations: Form, function, and technology' in S. Oskamp & S. Spacapan (eds) *Human reactions to technology: The Claremont symposium on applied social psychology* (Beverly Hills: Sage Publications).

A. Lashinsky (2012) *Inside apple: The secrets behind the past and future success of Steve Jobs's Iconic Brand* (London: John Murray).

B. Laurel (1993) *Computers as theatre* (Boston: Addison-Wesley).

P. R. Lawrence & J. W. Lorsch (1967) *Organization and environment: Managing differentiation and integration* (Harvard University, Graduate School of Business Administration).

M. Lepper & T. Malone (1987) 'Intrinsic motivation and instructional effectiveness in computer-based education' in R. Snow & M. Farr (eds) *Aptitude, learning and instruction*, vol. 3 (Hillsdale: Erlbaum).

S. Levy (2011) *In the plex: How Google thinks, works, and shapes our lives* (New York: Simon & Schuster).

M2 Research (2012) *Gamification in 2012*, http://gamingbusinessreview.com/wp-content/uploads/2012/05/Gamification-in-2012-M2R3.pdf.

C. Mainemelis & S. Ronson (2006) 'Ideas are born in fields of play: Towards a theory of play and creativity in organizational settings', *Research in Organizational Behavior*, 27, 81–131.

S. Malliet & G. De Meyer (2005) 'The history of the video game' in J. Raessens & J. Goldstein (eds) *Handbook of computer game studies* (Cambridge: The MIT Press).

I. S. Mayer (2009) 'The gaming of policy and the politics of gaming: A review', *Simulation & Gaming*, 40, 6, 825–862.

K. McCoy (2013) 'Apple CEO defends tax tactics at Senate hearing', *USA TODAY*, http://www.usatoday.com/story/money/business/2013/05/21/apple-tax-hearing/2344351/.

J. McGonigal (2011) *Reality is broken: Why games make us better and how they can change the world* (London: Jonathan Cape).

J. E. McGrath & A. B. Hollingshead (1994) *Groups interacting with technology* (Thousand Oaks: Sage Publications).

B. McLean & P. Elkind (2003) *The smartest guys in the room: The amazing rise and scandalous fall of Enron* (New York: Portfolio).

D. H. Meadows, Club of Rome, et al. (1972) *The limits to growth: A report for the Club of Rome's project on the predicament of mankind* (New York: Universe Books).

E. Medina (2005) 'Digital games: A motivational perspective', *Proceedings of DiGRA 2005 conference: Changing views – Worlds in play* (Vancouver).

J. P. Mestre (2005) (ed.) *Transfer of learning from a modern multidisciplinary perspective* (Greenwich: IAP).

R. E. Miles & C. C. Snow (1978) *Organizational strategy, structure, and process* (New York: McGraw-Hill).

J. Mingers & L. White (2010) 'A review of the recent contribution of systems thinking to operational research and management science', *European Journal of Operational Research*, 210, 1147–1161.

H. Mintzberg (1983) *Structure in fives: Designing effective organizations* (Englewood Cliffs: Prentice-Hall).

H. Mintzberg (1987) 'The strategy concept I: Five Ps for strategy' in G. R. Carroll & D. Vogel (eds) *Organizational approaches to strategy* (Cambridge: Ballinger).

J. Moisander & S. Stenfors (2009) 'Exploring the edges of theory-practice gap: Epistemic cultures in strategy-tool development and use', *Organization*, 16, 2, 227–247.

E. Mollick & N. Rothbard (2012) 'Mandatory fun: Gamification and the impact of games at work', *The Wharton School Research Paper*, 22.

C. Morningstar & R. Farmer (1991) 'The lessons of Lucasfilm's Habitat' in M. Benedikt (ed.) *Cyberspace: First steps* (Cambridge: MIT Press).

D. Nadler & M. L. Tuschman (1997) *Competing by design: The power of organizational architecture* (New York: Oxford University Press).

M. Neumeier (2009) *The designful company: How to build a culture of nonstop innovation* (Berkeley: New Riders).

R. Newton (2007) *Project management, step by step: How to plan and manage a highly successful project* (Harlow: Pearson Prentice Hall Business).

Newzoo (2012) *Infographic 2012 – US*, http://www.newzoo.com/infographics/infographic-2012-us/.

H. Owen (2008) *Open space technology: A user's guide*, 3rd ed. (San Francisco: Berret-Koehler).

T. J. Peters & R. H. Waterman (1982) *In search of excellence: Lessons from America's best-run companies* (New York: Harper & Row).

J. Pfeffer & G. R. Salancik (1978) *The external control of organizations: A resource dependence perspective* (New York: Harper & Row).

M. Pina e Cunha & J. Vieira da Cunha (2006) 'Towards a complexity theory of strategy', *Management Decision*, 44, 7, 839–850.

D. H. Pink (2009) *Drive: The surprising truth about what motivates us* (New York: Riverhead Books).

M. E. Porter (1980) *Competitive strategy* (New York: Free Press).

W. Poundstone (1992) *Prisoner's dilemma* (New York: Doubleday).

J. B. Quinn (1980) *Strategies for change, logical incrementalism* (Homewood: Irwin).

B. Reeves, T. W. Malone & T. O'Driscoll (2008) 'Leadership's online labs', *Harvard Business Review*, 86, 5, 58–66.

B. Reeves & J. L. Read (2009) *Total engagement: Using games and virtual worlds to change the way people work and businesses compete* (Boston: Harvard Business School Press).

P. Regnér (2003) 'Strategy creation in the periphery: Inductive versus deductive strategy making', *Journal of Management Studies*, 40, 1, 57–82.

M. Robertson (2010) 'Can't play, won't play', *Hide & Seek*, http://www.hideandseek.net/2010/10/06/cant-play-wont-play/.

A. Romme (2003) 'Making a difference: Organization as design', *Organization Science*, 14, 558–573.

A. Romme (2004) 'Perceptions of the value of microworld simulation: Research note', *Simulation & Gaming*, 35, 3, 427–436.

M. Rose (2012) 'World of Warcraft gains 1M subs following Pandaria release', *Gamasutra*, http://www.gamasutra.com/view/news/178844/World_of_Warcraft_gains_1M_subs_following_Pandaria_release.php.

Rovio (2011) *Angry Birds Smashes Half a Billion Downloads*, http://www.rovio.com/en/news/press-releases/95/angry-birds-smashes-half-a-billion-downloads.

Rovio (2013) *Rovio Entertainment to aunch One of the World's Biggest Video Networks in Its Games*, http://www.rovio.com/en/news/press-releases/276/rovio-entertainment-to-launch-one-of-the-world's-biggest-video-networks-in-its-games/.

R. M. Ryan & E. L. Deci (2000) 'Self-determination theory and the facilitation of intrinsic motivation, social development, and well-being', *American Psychologist*, 55, 1, 68–78.

R. M. Ryan, C. S. Rigby & A. Przybylski (2006) 'The motivational pull of video games: A self-determination theory approach', *Motivation and Emotion*, 30, 347–363.

W. A. Sahlman (2010) 'Management and the financial crisis', *Economics, Management, and Financial Markets*, 5, 4, 11–53.

K. Salen & F. Zimmerman (2004) *Rules of play: Game design fundamentals* (Cambridge: The MIT Press).

J. Schell (2008) *The art of game design: A book of lenses* (Burlington: Morgan Kaufmann).

D. J. Schiano, B. Nardi, T. Debeauvais, N. Ducheneaut & N. Yee (2011) 'A new look at World of Warcraft's social landscape', *6th International Conference on the foundations of digital games* (Bordeaux, France).

K. Schmidt (2002) 'The problem with awareness', *Computer Supported Cooperative Work*, 11, 285–298.

J. Schmitz & J. Fulk (1991) 'Organizational colleagues, media richness, and electronic mail', *Communication Research*, 18, 4, 487–523.

M. Scholten (2013) 'Boogerd bekent dan toch dopinggebruik', *NRC*, http://www.nrc.nl/nieuws/2013/03/06/boogerd-bekent-dan-toch-dopinggebruik/.

W. R. Scott (2008) *Institutions and organizations: Ideas and interests*, 3rd ed. (Los Angeles: Sage Publications).

W. R. Scott & G. F. Davis (2007) *Organizations and organizing: Rational, natural, and open systems perspectives* (Upper Saddle River: Pearson Prentice Hall).

J. Seely Brown & D. Thomas (2008) 'The gamer disposition', *Harvard Business Review*, 86, 2, 28.

R. Semler (2004) *The seven-day weekend: Changing the way work works* (New York: Portfolio).

D. W. Shaffer (2006) *How computer games help children learn* (New York: Palgrave Macmillan).

H. A. Simon (1996) *The sciences of the artificial*, 3rd ed. (Cambridge: The MIT Press).

S. J. Spear (2004) 'Learning to lead at Toyota', *Harvard Business Review*, 82, 5, 78–86.

C. Steinkuehler (2004) 'Learning in massively multiplayer online games' in Y. Kafai, W. Sandoval, N. Enyedy, A. Nixon & F. Herrera (eds) *Proceedings of the sixth international conference of the learning sciences* (Mahwah: Erlbaum).

C. Steinkuehler (2005) 'The new third place: Massively multiplayer online gaming in American youth culture', *Tidskrift for Lararutbildning och forskning*, 12, 3, 16–33.

J. B. Stewart (2009) 'Eight days: The battle to save the American financial system', *The New Yorker*, September 21, 2009, http://www.newyorker.com/reporting/2009/09/21/090921fa_fact_stewart.

A. Stoppelenburg, L. de Caluwé & J. Geurts (2012) *Gaming: Organisatieverandering met spelsimulaties* (Deventer: Kluwer).

Submarine channel (2011) 'Interview with game designer Eric Zimmerman', *Submarine Channel*, http://www.submarinechannel.com/profiles/eric-zimmerman/.

B. Suits (1978) *The grasshopper: Games, life, and Utopia* (Toronto: University of Toronto Press).

B. Sutton-Smith (1997) *The ambiguity of play* (Cambridge: Harvard University Press).

D. Tapscott & A. D. Williams (2006) *Wikinomics: How mass collaboration changes everything* (New York: Portfolio).

F. W. Taylor (1911) *The principles of scientific management* (New York: Harper & Brothers).

T. Taylor (2006) *Play between worlds: Exploring online game culture* (Cambridge: The MIT Press).

K. Tee, S. Greenberg & C. Gutwin (2006) 'Providing artifact awareness to a distributed group through screen sharing', in P. Hinds & D. Martin (eds) *Proceedings of the conference on computer supported cooperative work (CSCW 2006)* (Banff: ACM).

C. Thompson (2007) 'Halo 3: How microsoft labs invented a new science of play', *Wired Magazine*, http://www.wired.com/gaming/virtualworlds/magazine/15-09-ff_halo.

S. Turkle (1995) *Life on the screen: Identity in the age of the internet* (New York: Simon & Schuster).

E. Vaara & R. Whittington (2012) 'Strategy-as-practice: Taking social practices seriously', *The Academy of Management Annals*, 6, 1, 285–336.

J. E. van Aken (2004) 'Management research based on the paradigm of the design sciences: The quest for field-tested and grounded technological rules', *Journal of Management Studies*, 41, 2, 219–246.

J. E. van Aken (2007) 'Design science and organizational development interventions: Aligning business and humanistic values', *The Journal of Applied Behavioral Science*, 43, 1, 67–88.

J. van Bree (2013) *Exploring the rules by playing the game: An investigation of the potential of video game design to enrich organizations* (Breukelen: Nyenrode Business Universiteit).

J. van Bree, M. Copier & T. Gaanderse (2010) 'Designing an organizational rule set: Experiences of using second-order organizational design in healthcare', *International Journal for Organisational Design and Engineering*, 1, 1/2, 29–54.

J. van Bree, S. de Lat (2011) 'Complex systems and emergent behavior: Engaging with computer games to enrich organization studies', *Nyenrode Research Paper No. 11-05*, http://ssrn.com/abstract=1959369.

A. H. Van de Ven, M. Ganco & C. R. Hinings (2013) 'Returning to the frontier of contingency theory of organizational and institutional designs', *The Academy of Management Annals*, 7, 1, 391–438.

J. van Mastrigt (2006) The Big Bang (presentation), at: *Dutch Gamedays* (Utrecht).

VentureBeat (2013) *Marketing, Reviews, and Mobile May All Determine If Your Game Is a Hit or a Flop*, http://venturebeat.com/2013/03/28/eedar-game-data/.

J. Von Neumann & O. Morgenstern (1944) *Theory of games and economic behavior* (Princeton: Princeton University Press).

G. Walker (2009) *Modern competitive strategy*, 3rd ed. (Boston: McGraw-Hill).

J. P. Walther (1996) 'Computer-mediated communication: Impersonal, interpersonal and hyperpersonal interaction', *Communications Research*, 23, 1, 3–43.

H. Warmelink (2013) *Towards playful organizations: How online gamers organize themselves (and what other organizations can learn from them)* (Delft: Next Generation Infrastructure Foundation).

K. E. Weick (2004) 'Rethinking organizational design' in R.J. Boland & F. Collopy (eds) *Managing as designing* (Stanford: Stanford University Press).

M. R. Weisbord & S. Janoff (2010) *Future search: Getting the whole system in the room for vision, commitment, and action*, 3rd ed. (San Francisco: Berrett-Koehler).

J. R. Whitson (2010) 'Rule making and rule breaking: Game development and the governance of emergent behaviour', The Fibreculture Journal, http://sixteen.fibreculturejournal.org/rule-making-and-rule-breaking-game-development-and-the-governance-of-emergent-behaviour/.

D. Williams, N. Ducheneaut, L. X. Zhang, N. Yee & E. Nickell (2006) 'From tree house to barracks: The social life of guilds in World of Warcraft', *Games and Culture*, 1, 4, 338–361.

D. W. Winnicott (1971) *Playing and reality* (New York: Basic Books).

K. D. Wolf (2007) 'Communities of practice in MMORPGs: An entry point into addiction' in C. Steinfeld, B. T. Pentland, M. Ackerman & N. Contractor (eds) Communities and technologies 2007: *Proceedings of the third communities and technologies conference* (London: Springer-Verlag).

R. E. Wright (2010) 'Teaching history in business schools: An insider's view', *Academy of Management Learning & Education*, 9, 4, 697–700.

E. Zimmerman (2012) 'Jerked around by the magic circle – Clearing the air ten years later', *Gamasutra*, http://www.gamasutra.com/view/feature/135063/jerked_around_by_the_magic_circle_.php.

Index

GPSR Compliance
The European Union's (EU) General Product Safety Regulation (GPSR) is a set
of rules that requires consumer products to be safe and our obligations to
ensure this.

If you have any concerns about our products, you can contact us on

ProductSafety@springernature.com

In case Publisher is established outside the EU, the EU authorized
representative is:

Springer Nature Customer Service Center GmbH
Europaplatz 3
69115 Heidelberg, Germany